2nd International Conference on Composite Material, Polymer Science and Engineering (CMPSE 2018)

MATEC Web of Conferences Volume 264 (2018)

Osaka, Japan
21 - 22 September 2018

Editor:

Katsuyuki Kida

ISBN: 978-1-5108-8074-0

Printed from e-media with permission by:

Curran Associates, Inc.
57 Morehouse Lane
Red Hook, NY 12571

Some format issues inherent in the e-media version may also appear in this print version.

For additional information, please contact EDP Sciences – Web of Conferences
at the address below.

EDP Sciences – Web of Conferences
17, Avenue du Hoggar
Parc d'Activité de Courtabœuf
BP 112
F-91944 Les Ulis Cedex A
France

Phone: +33 (0) 1 69 18 75 75
Fax: +33 (0) 1 69 28 84 91

contact-edps@webofconferences.org

Additional copies of this publication are available from:

Curran Associates, Inc.
57 Morehouse Lane
Red Hook, NY 12571 USA
Phone: 845-758-0400
Fax: 845-758-2633
Email: curran@proceedings.com
Web: www.proceedings.com

TABLE OF CONTENTS

The International Conference on Composite Material, Polymer Science and Engineering

(CMPSE2018）

Sep. 21-22, 2018

Osaka Japan

Edited by

Katsuyuki Kida, Professor, University of Toyama, Japan

CMPSE2018 Committee

General Chair

Professor Katsuyuki Kida, University of Toyama, Japan Scientific Editor

Program Committee

Seungwon Lee, University of Toyama, Japan

Koshiro Mizobe, University of Toyama, Japan

Takahiro Matsueda, Tokyo Metropolitan University, Japan

Petre ANGHELESCU, University of Pitesti, Romania

Claudia Canali, University of Modena and Reggio Emilia, Italy

Bhingole Pramod P., Institute of Infrastructure, Technology, Research and Management (IITRAM), India

Jaspal Singh, Punjab Agricultural University, India

Toby D. Young, Institute of Fundamental Technological Research of the Polish Academy of Sciencesul, Poland

Abdul Talib bin Din, Universiti Teknikal Malaysia Melaka(UTeM), Malaysia

Zakiah Ahmad, University Teknologi Mara, Malaysia

Mohd Yuhazri Bin Yaakob, Universiti Teknikal Malaysia Melaka(UTeM), Malaysia

Hariati Taib, Universiti Tun Hussein Onn Malaysia, Malaysia

Dmitry A. Chinakhov, Yurga Institute of Technology of National Research Tomsk Polytechnic University, Russia

Valentino Paradiso, University of Salerno– Faculty of Engineering, Italy

Muhammad Naufal Bin Mansor, University of Malaysia Perlis (UniMAP)

Yasin Polat, Nevşehir Hacı Bektaş Veli University, Turkey

Seyed Saeid Rahimian Koloor, Universiti Teknologi Malaysia (UTM), Malaysia

Academic Members

Ernie Suzana Binti Ali, Universiti Sains Islam Malaysia, Malaysia

Mimi Azlina Abu Bakar, University Technology MARA, Malaysia

Mohd Rasidi Bin Ibrahim, Universiti Tun Hussein Onn Malaysia, Malaysia

Shahril Anuar Bin Bahari, Universiti Teknologi MARA (UiTM) Malaysia, Malaysia

Saraev Yu. N. ,Institute of Strength Physics and Materials Science of Siberian, Russia

Carlos Henrique Lemos Soares, University of Campinas, SP, Brazil

Lai Chin Wei, University of Malaya, Malaysia

Ervina Efzan Mhd Noor, Multimedia University, Malacca, Malaysia

Muhammad Iqbal Hussain, Universiti Malaysia Perlis (UniMAP), Malaysia

Mohd Zaidi Sidek, University Malaysia Pahang (UMP), Malaysia

Nazuhusna Binti Khalid, Universiti Malaysia Perlis (UniMAP), Malaysia

M T Prathap Kumar, RNS Institute of Technology, India

Hab. Artur Chrobak, University Of Silesia, Institute of Physics, Poland

Mohamad Bin Awang, Universiti Malaysia Terengganu (UMT), Malaysia

Melchor J. Potestas, Mindanao State University – Iligan Institute of Technology, Philippines

Hugo Miguel Andrade Lopes Figueiredo da Silva, University of Minho, Portugal

Preface

This proceeding book consists of 18 papers from 'The 2nd International Conference on Composite Material Polymer Science and Engineering (CMPSE)', which was held on September 21st and 22nd, 2018 in Osaka, Japan. This conference focused on recent scientific research and engineering technology in the field of composite materials, polymer science and engineering. The main topics of the 2nd CMPSE were 'Mechanical properties of corrosion, damage analysis, strength and wear', 'Thermal properties of thermal stability and thermo-plasticity' and 'Product manufacturing of recycle, material process and machining. All papers were accepted through a peer-review process by the editorial committee and international reviewers, and were categorized into 3 chapters.

On behalf of the organizing committee, the editor would like to acknowledge the technical comments from the reviewers and thank all of the authors for their contributions to this proceeding book. And thanks are expressed to Koshiro Mizobe from the University of Toyama, who assigned the conference secretary thorough the planning of the conference, and Kazuaki Nakane from Osaka University and Hitonobu Koike from Miyazaki University for their editorial work.

Unfortunately, typhoon Jebi (No. 21) hit Osaka area and destroyed the main bridge between Kansai floating airport and the main land. While many researchers and professors canceled the conference, over 30 people attended. On behalf of the committee, the editor would appreciate their support of the conference.

Editor

Katsuyuki Kida, Professor, University of Toyama, Japan

December 28, 2018

Statement of Peer review

In submitting conference proceedings to *Web of Conferences*, the editors of the proceedings certify to the Publisher that

1. They adhere to its **Policy on Publishing Integrity** in order to safeguard good scientific practice in publishing.
2. All articles have been subjected to peer review administered by the proceedings editors.
3. Reviews have been conducted by expert referees, who have been requested to provide unbiased and constructive comments aimed, whenever possible, at improving the work.
4. Proceedings editors have taken all reasonable steps to ensure the quality of the materials they publish and their decision to accept or reject a paper for publication has been based only on the merits of the work and the relevance to the journal.

Title, date and place of the conference

2nd International Conference on Composite Material, Polymer Science and Engineering, Sept. 21-22, 2018, Osaka, Japan

Proceedings editor(s):

Date and editor's signature

Dec. 28, 2018

Fatigue Damage Analysis of Composite Insulator Cor Rod/sheath Interface

Xinze Zhao, Senming Xi , Meiyun Zhao[a], Wenyun PENG, Seng Tian, Wei Li

Hubei Key Laboratory of Hydroelectric Machinery Design & Maintenance, China Three Gorges University, Yichang, 443002

Abstract--A self-made fatigue testing machine was used to simulate the damage process of the insulator interface under working conditions and the damage degree is evaluated by the compressive shear test. The improved interface adhesion model reflects the interface debonding process under different fatigue loads by introducing the damage coefficient χ, migration coefficient α and the debonding coefficient μ. The relationship of fatigue loads and interface damage is established by the peeling force, the unit interface separation work and the peeling time. The test demonstrates that the interface damage near the bottom of the insulator is worse than that near the middle part. The result of this analysis will be beneficial in easily finding damaged places of the insulator, and so as to reduce the power breakdown.

1 Introduction

The composite insulator is an insulating device widely used in high voltage transmission line, mainly consisting of metal fittings, core rod, and housing. The core rod consists of fiber-glass and epoxy resin, and it is the main load-bearing part. The main material of the sheath is silicone rubber. Cohesive interface between core rod and sheath runs throughout two ends of metal fittings, which is the key part of the insulation [1-2]. Due to the effects of the overhead line weight, the wind load and other factors, the force of composite insulator in operation is very complicated [3-5]. The mechanical fatigue, environment temperature changing, atmosphere corrosion, high field strength and so on, all lead to core rod/sheath interface deteriorative failure [6-8], and eventually may result to the serious accidents [9-11]. It is of great significance to analyze the fatigue damage of the bonding interface of the core rod and sheath under different loads, which can reduce the failure of insulators. At home and abroad, the interface is evaluated by measuring the interfacial bonding strength or detecting interface defects. You et al. [12] introduced a method to test the interfacial bonding strength of two sides of the flexible matrix with a 180-degree angle stripping. Zhang et al. [13] used the interface constitutive model and 90-degree peeling test to analyze the bonding strength of the anisotropic conductive adhesive film from theoretically and experimentally, and used the damage factors to illustrate the changes of interface bonding strength under high temperature and humidity loading. The interfacial bonding strength was divided into several grades by the anatomical grading test method, and the improved water diffusion experiment could discover the quality defect of the composite insulator that traditional methods could not find [14-15]. Duan et al. [16] used the core-sheath compressive shear test to evaluated the interfacial adhesion strengths of the composite insulator. Most of them used the confirmatory test to measure the interfacial quality between sheath and core, while the fatigue parameters are not considered, it is difficult to indicate the influence about degree of damage to the interfacial adhesion of the insulator.

In this paper, we combined the constitutive model of bonding interface proposed by Needleman. by means of theoretical analysis and compressive shear test, the peeling process of core rod/sheath bonding interface under the different fatigue conditions was analyzed and the change of residual bond strength of the interface was described with the unit area separation work and the peeling time. It provides a reference to the research of the mechanical fatigue failure of composite insulators.

2 Test

2.1 Core rod/ sheath interface and parameters

FXBW4-110/100 suspension composite insulator which is used in 110kV transmission line, was selected as the test object. Under normal circumstances the interface between sheath and core rod should be tightly bonded. However, due to the mechanical fatigue produced by self-weight of transmission line, swing and torsion under wind load [17-18], the interface of composite insulator could have air gaps and even crack. Under the working conditions the force of the composite insulator core rod/sheath interface is shown in Figure 1.

[a] Corresponding author: zhaomeiyun@ctgu.edu.cn

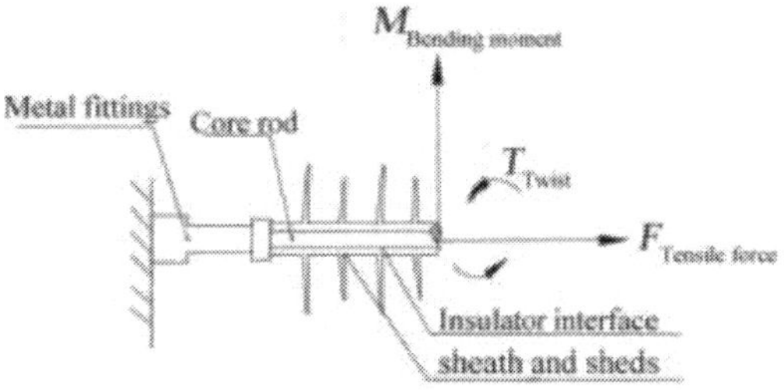

Figure 1. Force diagram of composite insulators on core/sheath interface

2.2 Fatigue test

In order to simulate the fatigue failure process of core rod/sheath interface of composite insulators, dynamic fatigue test was carried out on a self-made fatigue device until the insulator being ruptured, as shown in Figure 2 [19]. The test machine can put different frequency pull, bend, torsion load on the composite insulator to simulate the process of fatigue under the wire weight, and wind load including swing and twist. Insulator fatigue test system is shown in Figure 3, and the test design is shown in Table 1.

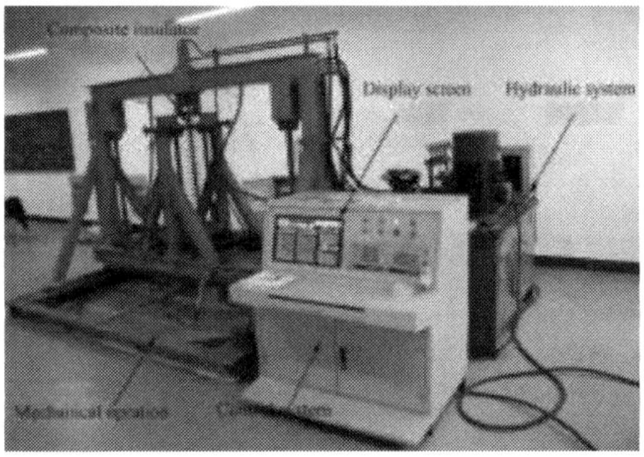

Figure 2. Fatigue test machine

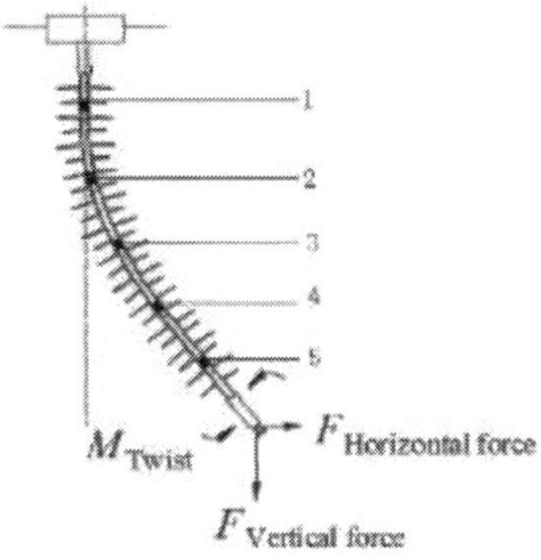

Figure 3. Composite insulator force diagram

Table. 1 Fatigue test scheme of composite insulators

Test groups	Tensile force F/MPa	Swing amplitude L/mm	Twist angle α/ °
1	2.7	380	5
2	2.7	380	10
3	2.7	380	15
4	2.7	380	20
5	2.7	380	25
6	2.7	130	30
7	2.7	180	30
8	2.7	230	30
9	2.7	280	30
10	2.7	330	30
11	2.7	380	30

2.3 Peel test

Because the defects of the composite insulator are generally located internal, and it is difficult to detect [20]. In this paper, A device was designed to measure the interfacial adhesive strength between the sheath and core rod, the interfacial adhesive strength can reflect the degree of damage, as shown in Figure 4. Combining with a universal pressure test machine, then the peeling force, displacement and peeling time can be recorded.

Figure 4. The equipment of peeling the sheath

Take the fatigue test insulators as the specimens. Every insulator is divided into five equal parts, each of which is 200 mm, and then intercept the specimen which the length of sheath is 30 mm, length of the core rod is 50 mm from each part, and number 1-5, as shown in Figure 3. In order to avoid the influence of radial elastic deformation of the sheath when the sheath is being peeled off, the sheath of the specimen is divided in advance into four equal sections in the circumferential direction. Figure 5 presents the peeling force and displacement curve of interface debonding process.

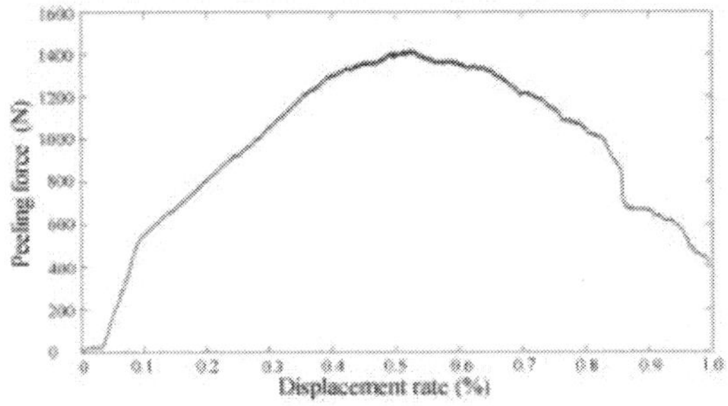

Figure 5. The curve of interface debonding process

3 Peering mechanics analysis

The constitutive model of the bonded region established by Needleman et al. describes the continuous relationship between the tensile and the interfacial debonding displacement of unit area, without regard to the nature of the bonding material and the bonded material as a direct influencing factor. The vertical and shear displacements between two points of the material bonding interface can be expressed as:

$$\begin{cases} \boldsymbol{T_n} = \boldsymbol{n} \cdot \boldsymbol{T} \\ \boldsymbol{T_b} = \boldsymbol{b} \cdot \boldsymbol{T} \\ \boldsymbol{T_t} = \boldsymbol{t} \cdot \boldsymbol{T} \end{cases} \quad (1)$$

Where n、b、t are the three directions of the Cartesian coordinate system, $\boldsymbol{T}$ is the unit vector in three direction, and T_n, T_b, T_t are the separation forces normalized for the interface and tangent to the interface at the interface crack tip per unit area.

Zhang [22] thinks that there is a linear relationship between the tension and the displacement along the interface and the tension-displacement along the vertical interface according to the Needleman's interfacial adhesion model. The peeled off the interface crack tip bonding point is taken as point a, and the initial peeling point on the sheath is taken as the b, so these two points along the axial displacement is u_b, and displacement perpendicular to the interface is u_n. When the sheath detached from the core rod completely, the displacement vertical to interface u_n set to δ, which represents the interface sticking feature length [21-22].

$$\begin{cases} \upsilon_n = \dfrac{u_n}{\delta} \\ \upsilon_b = \dfrac{u_b}{\delta} \\ \upsilon_t = \dfrac{u_t}{\delta} \end{cases} \quad (2)$$

$$T_n = E \cdot F(d) \cdot \upsilon_n \qquad u_n \le \delta \quad (3)$$

$$T_b = G_b \cdot F(d) \cdot \upsilon_b \qquad u_b \le \delta \quad (4)$$

$$T_t = G_t \cdot F(d) \cdot \upsilon_t \qquad u_t \le \delta \quad (5)$$

On basis of the model established by Needleman, Tvergaard [23] limited:

$$F(d) = (1-d)^2 \quad (6)$$

$$d = \max \sqrt{\upsilon_n^{\ 2} + \upsilon_b^{\ 2} + \upsilon_t^{\ 2}} \quad (7)$$

The potential energy expression of the bonding interface is as follows:

$$\phi sep(u_n, u_b) = -\int_0^u [T_n du_n + T_b du_b] \quad (8)$$

The relationship between the theoretical peeled tensile force and displacement is derived from the constitutive relationship between the viscoelastic body and the bonding interface.

Combined with the tool peeling force model in [24] and the structural characteristics of the composite insulator core rod and sheath, when the interface between sheath and core rod is peeled completely and the tool is wedged into the interface, the inclination of the tool bears partially the pressure F_{N1} of the sheath, friction F_{f1} along the sheath tangential, core rod pressure F_{N2} on the tool, along the core rod tangential friction F_{f2}. The tool force is shown in Figure 6 (a). Corresponding, the peeling pressure of tool to the bonding interface causes it to separate gradually, and the force is decomposed into an opening force T_n perpendicular to the interface and a shear force T_b along the interface. The force separated at the crack tip of the bonding interface is shown in Figure 6 (b).

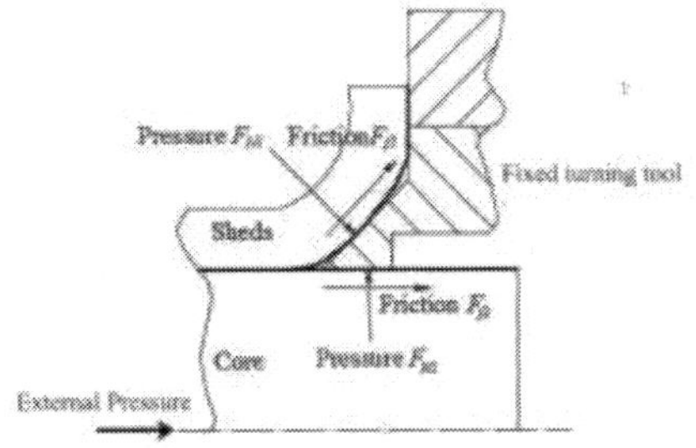

(a) The Tool-stressed model

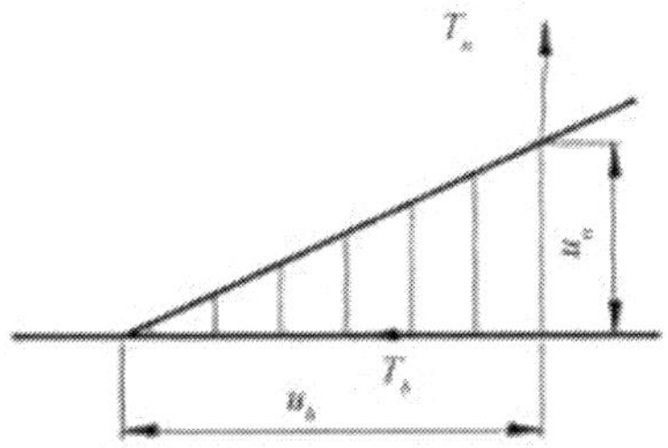

(b) The force separated at the crack tip of the bonding interface

Figure 6. The schematic of the separation of the core/sheath interface

The peel test shows that u_n is along the vertical interface, u_b is perpendicular to the interface, and the difference between them is very small, so we can assume

https://doi.org/10.1051/matecconf/201926401001

that $u_n = u_b$. Due to the extrusion of the core rod on the inner wall of the annular tool, the pressure still exists after debonding the sheath completely. According to the formula (3)-(8), the formula is amended as:

$$T_b = G\cdot(\upsilon_b^{3} - 2\sqrt{2}\cdot\upsilon_b^{2} + 2\upsilon_b) \tag{9}$$

According to the stress formula of shear test,

$$\tau = \frac{F}{S\cdot\upsilon_b} \tag{10}$$

Where τ is the interfacial shear strength, F is the force from press machine, and S is the bonded interface area of the specimen. Combine formula (9) and (10), get the relationship of pressure shear force and displacement as follows.

$$F = G\cdot S\cdot(\upsilon_b^{4} - 2\sqrt{2}\cdot\upsilon_b^{3} + 2\cdot\upsilon_b^{2}) \tag{11}$$

After fatigue dynamic load test, the interface between housing and core rod of the insulators would occur damage inevitably, which leads to the interfacial bond strength decreased. In order to show the constitutive relation of the interface after the fatigue damage, the coefficient χ, the migration coefficient α and the debonding coefficient μ are introduced. The coefficient χ is the degree of pressure drop in the process of peeling and α is the change of the position of maximum peeling force F_{max}, and the debonding coefficient μ describes the slope of the peeling force and displacement when the interface debonding finally. After the coefficients being corrected, the interface damage model of core rod/sheath considering fatigue influence is obtained as follows:

$$F = \chi\cdot G\cdot S\cdot\left\{(\upsilon_b+\alpha)^{4} - 2\sqrt{2}\cdot(\upsilon_b+\alpha)^{3} + \mu\cdot(\upsilon_b+\alpha)^{2}\right\} \tag{12}$$

4 Results and discussion

4.1 Peeling force - displacement relationship

Here we aim to analyse the damage degree of the adhesive interface of the same position of the different insulator specimens under the situation, that the tensile force and torsion angle in experiment remain constant but swing amplitude is variable. The self-made pressure-shear peeling device is used to separate the core rod-sheath interface. The compressive shear device is used to measure the interface quality. The shear force-displacement curve was obtained as shown in Figure 7. and the modified theoretical model of interfacial adhesion damage was verified reasonable.

Figure 7 indicates the experimental values curve can reflect the change of interface peel test, but the theoretical values of peel force is lower than the experimental value in the initial stage. The main reason is the material elastic deforms leading to the interface suffer the resistance at this stage and it results in deviation between the actual peel force and theoretical value. Table 2 shows the change of fatigue damage factor under different swing.

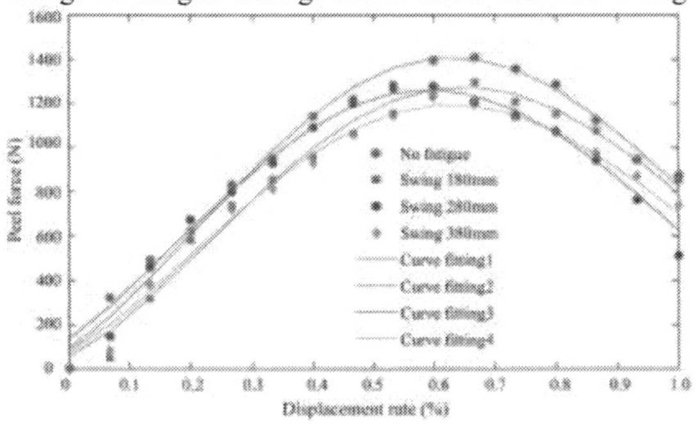

(a) The sampling point No.1

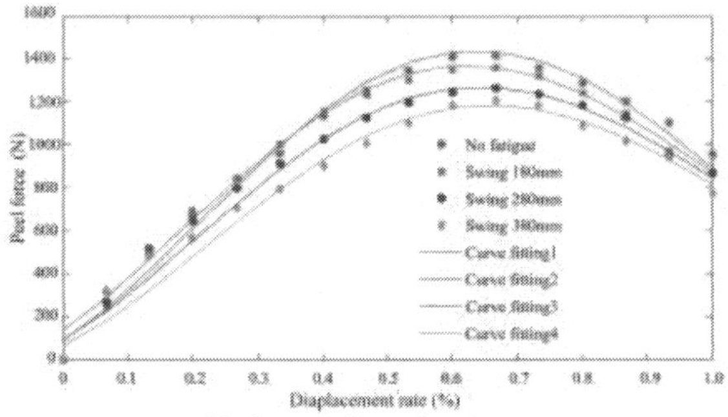

(b) The sampling point No.5

Figure 7. Change curves of peel force with displacement rate

Table 2. Fatigue damage factor under different swing

Swing amplitude *L*/mm	Sampling point No.1			Sampling point No.5		
	coefficient χ	Debonding coefficient μ	Migration coefficient α	coefficient χ	Debonding coefficient μ	Migration coefficient α
0	1.302	2.03	0.1	1.302	2.03	0.1
180	1.191	2.03	0.08	1.192	2.07	0.13
280	1.179	2.03	0.13	1.105	2.07	0.11
380	1.105	2.03	0.1	1.031	2.07	0.095

As a result of the fatigue damage, the overall interface peel force decreases, which has no significant effect on the maximum peeling force F_{max}. As can be seen from Table 2, at specimen part 1, the damage factor χ decreases from 1.302 at no swing to 1.105 at 380 mm swing, decreasing by 0.197. The coefficient χ at sampling point 5 near the bottom of the insulator decreases from 1.302 without fatigue down to 1.031 at the swing of 380 mm, decreasing by 0.271. The coefficient χ changed more obviously on the specimen No. 5, while the displacement coefficient α showed no obvious change, and still fluctuating around 0.1, which means the stress on the core rod/sheath interface near the bottom is greater and the interface damage is more serious. And the change of fatigue conditions has no obvious influence on the location of the maximum of peeling force F_{max} appears. The locations where the maximum peeling force F_{max} all lie in about 2/3 of the total displacement on samples.

For No. 1 specimen near the upper metal fitting, When the fatigue load was applied, no matter how the swing changed, the debonding coefficient μ would not change any more and keep at 2.03. Meanwhile, the specimen No.5 near the lower metal fitting, with the swing amplitude increasing, the coefficient μ gradually increased, and eventually stay at 2.07. It presents that the location of the force to the insulator has an important influence on the interface debonding coefficient μ, but the swing amplitude has little effect on it.

4.2 Interface separation work analysis

Integrating the load-displacement curve of the compressive shear test and using the unit area separation work that sheath peeled away from the core rod can analyze interface bond strength changes intuitively. Figure 8 presents the unit area separation work of the core rod/sheath interface in different work conditions obtained by the compressive shear test.

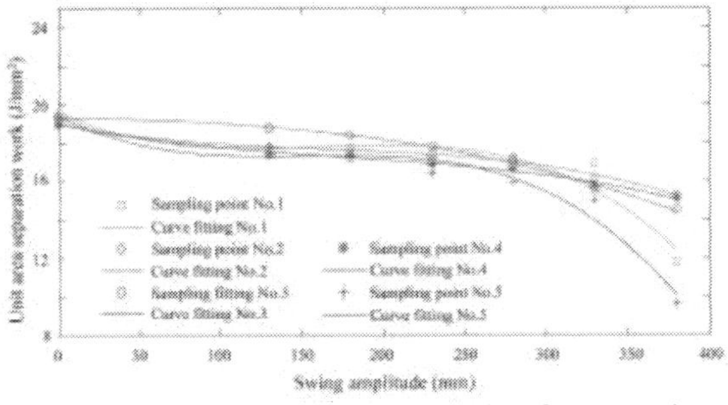

(a) The W-L curve of swing amplitude and interface separation work

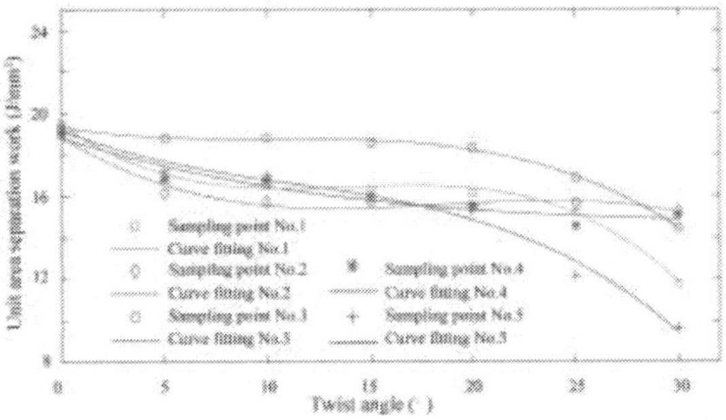

(b) The W-α curve of twist angle and interface separation work

Figure 8. Unit area interface separation work of various parts

As shown in Figure 8(a), when the swing is constant, the unit area separation work of all insulator specimens declines the increase of torsion angle. However, the change of the No.2 and No.4 specimens is little and the others decline rapidly once the twist angle is more than 20 degrees. That demonstrates the interface damage of corresponding parts are more serious. As is shown in Figure 8(b), when twisting angle is constant, the unit area separation work of interface of all insulator specimen is decrease with the increase of the swing. However, the change of the No.1 and No.5 specimen is large once the swing is more than about 250 mm and the decline of others is not obvious which shows the interface damage of the top and bottom parts are more serious. Then we can find the vulnerable locations of the composite insulator in operating conditions.

4.3 Interface peeling time

The compressive shear test was carried out by driving the press with constant power. The peeling time shows the sheath can reflect the strength of residual adhesion of the interface after the dynamic fatigue test. Figure 9 shows the time of sheath peeling from different parts of the insulator under the different fatigue parameters.

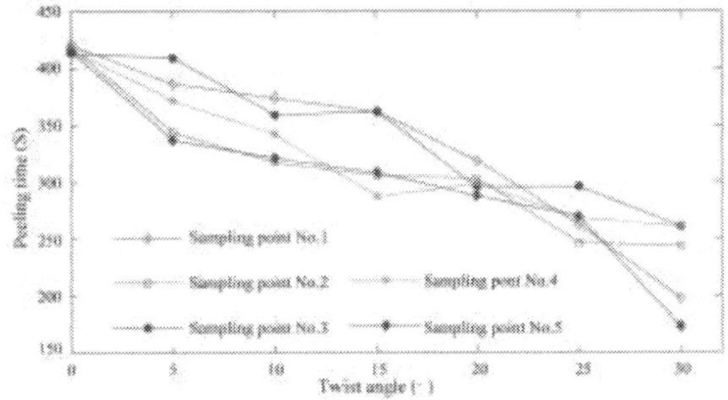

(a) Peeling time of samples in different twist angles

[a] Corresponding author: zhaomeiyun@ctgu.edu.cn

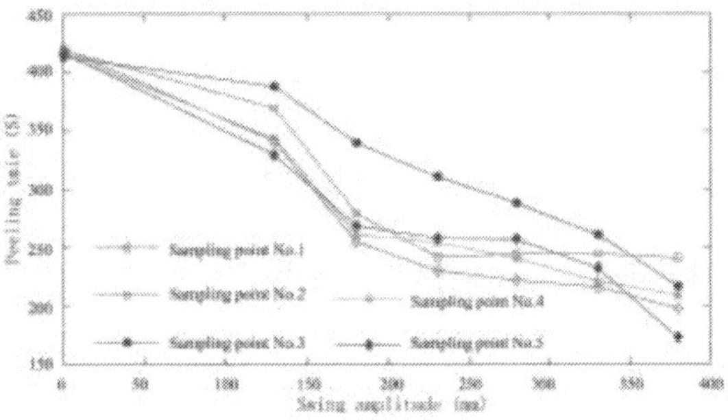

(b) Peeling time of samples in different swings
Figure 9. Peeling time of various location samples of the composite insulator

As shown in Figure 9(a), with the twist angle increasing especially with No.1 and No.5 specimens, the peeling time of the sheath away from the core rod is decreasing monotonically. Meanwhile, when the swing amplitude L is a variable, the peeling time of each part is more relevant with swings. For example, when the swing amplitude increases, except the peeling time of the specimen point No.3 which is located in the middle of the insulator decreases linearly, and the other parts go through the trend of firstly is quick then become slow and finally become quick. For the specimens point No.1 and No. 5, which is near two metal fittings, the peeling time declines most obviously. The results show that the change of the swing amplitude has little influence on the interface damage, but the interface near the joint of two metal fittings has serious damage. The change of torsion angle makes more damage to the insulator core rod/sheath interface.

5 Conclusions

We have successfully modified cohesive interfacial damage constitutive model and used it to characterize the process of core rod/sheath interface of the composite insulators under the different swing amplitudes and twist angles. We designed and made a device of peeling the sheath to analyse the damage degree of the interface.

We studied that the interface damage of the core rod/sheath when the swing amplitude and the twist angle are taken as the variable respectively. We confirm that the interface damage was affected by swing amplitude and twist angle and twist angles has more serious influence to the core rod/sheath interface than swing amplitude. Also, the peering time and force demonstrate that the damage of the two ends of the insulator near metal fitting is more serious than that in middle of the insulator under the same conditions.

Acknowledgement

This work was supported by the National Natural Science Foundation of China (No. 51475264 and 51605254); and Natural Science Foundation of Hubei Province (No. 2016CFB565).

References

1. T. YUAN, R. ZHANG, Q. ZHANG, et al., Insulators and Surge Arresters,138,19-24+31(2015)
2. Y.F. GAO, X.D. LIANG, Proceedings of the CSEE, 36, 5070-5077(2016)
3. F.Z. ZHANG, L. SONG, R.H. LI, et al., High Voltage Engineering, 38, 3093-3100 (2012)
4. W. HU, C. ZHAO, High Voltage Apparatus, 54, 93-97-103 (2018)
5. Q. XIAO, J.S. CAI, F. GAO, et al., Water Resources and Power, 29, 155-157+177 (2011)
6. X.Y. LI, X.D HAN, F.L ZHANG, et al., Insulators and Surge Arresters, 135, 7-13 (2012)
7. Z.S. XIE, M.S. Thesis, Dept. Mechanical Eng, North China Electric Power Univ., Heibei, China, (2012)
8. Z.S. XIE, Z.Q. WANG, Y. CHEN, et al., Transaction of China Electrotechnical Society, 29, 296-302(2014)
9. Y.B. SHAO, M. LU, D.Y. MA, et al., Insulators and Surge Arresters, 139, 34-39 (2016)
10. Q.G. ZHANG, J.B. ZHOU, H.B. XI, et al., High Voltage Engineering, 40, 987 -994 (2014)
11. F.C. LV, J.Q. MA, F.C. WANG, et al., High Voltage Engineering, 40, 3305-3310 (2014)
12. M. YOU, X. ZHENG, Analysis and application of adhesive strength, 23-58. (2009)
13. J. ZHANG, X. CHEN, H. JIA, Journal of Zheng Zhou University (Engineering Science), 27, 48-51 (2006)
14. Z.Y. SU, G. CHEN, Q.F. LI, et al., Power System Technology, 30, 53-57 (2006)
15. Y. CHEN, Y.S. LIU, J. SHEN, et al., Power System Technology, 30, 58-63 (2006)
16. M.M. DUAN, C. WANG, Y P. TU, et al., High Voltage Apparatus, 52,132-137 (2016)
17. F C. SUN, X D. LIANG, Y X. ZHOU, et al., Power System Technology. 26, 33-36 (2002)
18. M.Z. GAO, R. ZHANG, Z.W. ZHANG. High Voltage Engineering, 35,2591-2596. (2009)
19. X.Z. ZHAO, K WANG, Z.C. FU, et al., Dynamic torsional vibration testing machine for the composite insulators, China patent, 104833469B, (2017)
20. Z.L. HE, Ultrasonic detection of internal artificial defects in a composite insulator. South China University of Technology, China, (2013)
21. A. Needleman, Journal of Applied Mechanics, 54, 525~531 (1987)
22. J. ZHANG. Ph.D. dissertation, Dept. Chemical Engineering and Technology, Tianjin University, Tianjin, China, (2005)
23. V. Tvergaard, Material Science, 125, 203-213 (1990)
24. W.J. LIU, M.S. Thesis, Dept. Mechanical manufacturing and automation, Dalian University of Technology, Liaoning, China (2015)

Strength of composite plate with parallel offset misalignment double-bolted joint under bending moment

Narongkorn Krajangsawasdi[1, a] and Phacharaporn Bunyawanichakul[1]

[1]*Design Clinic Research Unit, Department of Aerospace Engineering, Faculty of Engineering, Kasetsart University, 50 Ngam Wong Wan Rd., Ladyaow, Chatuchak, Bangkok, 10900, Thailand*

Abstract. The connection concept of placing only two bolts in offset misalignment against the bending load along the wing span was used for an aerobatic airplane designed in Thailand as a KIT plane to minimize the impact of drilling numerous holes. This concept can deviate the force direction on the holes. The two suitable drilling positions should have the lowest resultant force and highest strength of fiber reinforcement structure. To investigate the strength of the fastener hole when the force deviated from the original orientation, specimens, made from twill weave carbon fiber-epoxy and laid up at ±45-degree orientation, were tested under bearing load according to ASTM D5961 standard. The experimental results revealed that the bearing strength of CFRP material decreases when the force deviation angle increases, so zero-angle deviation of the resultant force on the drilling hole is the most suitable orientation to absorb bolt bearing load. The most suitable pattern of two offset misalignment holes is the greatest horizontal distance at zero vertical distance when it was considered only the effect of the bearing strength and the deviation angle. Moreover, the failure pattern begins to deviate along the fiber orientation when the inclination angle increased.

1 Introduction

The parallel offset misalignment bolt joint is an interesting concept. It is applied to some KIT planes fixture such as Lancair Legacy (Figure 1) [1] and FURIO (Figure 2) [2]. The KIT plane concept allows customers of KIT planes or Experimental Light Sport Aircraft to assemble more than 50 percent of the overall aircraft components themselves. This facilitates delivery to buyers by separate parts, especially such large parts as the wing spar which is the focus of this paper. The concept considered here is called Torsional Resistance of the Bolt Group, and it uses two bolts to connect the left and right wings. The concept can counter the bending moment that plays an important role in the lift force on the wing. Every structure is always weakened by drilling holes, especially composite structure which is composed of a fiber fabric and resin matrix. Its strength depends on the direction of the fiber fabric. Therefore, the minimum number of bolts is two, this is considered suitable and safe regarding the fabric structure and drilling holes and supporting the wing load. The bolt fixing pattern and the load action on the structure considered in the current paper is illustrated in Figure 3.

There are three primary failure modes in a plate with a fastener hole when a bearing load is applied: local bearing, net tension, and shear out [3]. Figure 4 illustrates the different failure modes for the three failure patterns. It can be seen that the local bearing is the only a pattern which cannot break the whole structure, so it is defined as non-catastrophic failure, whereas the others are in the catastrophic group. The graph between load and deflection of a hole can be plotted in order to indicate the difference load absorption between catastrophic and non-catastrophic.

Figure 1. Lancair Legacy [1].

Figure 2. Unassembled FURIO 2 [2].

Figure 3. Double bolted fixed pattern with bending load generated by the vertical force.

[a] Corresponding author: narongkorn.k@ku.th

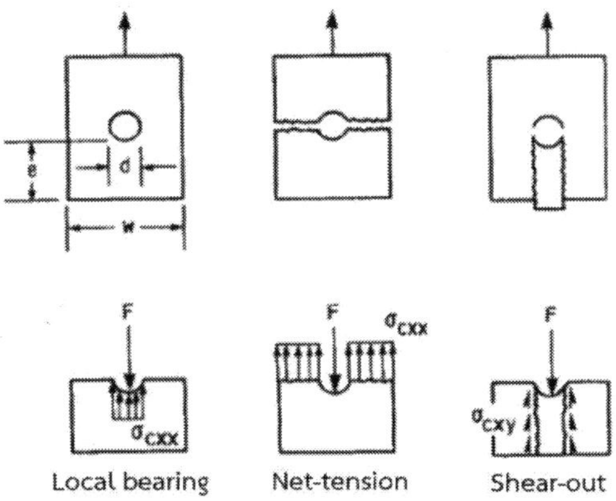

Figure 4. Three major failure modes of plate with a fastener subject to bearing load [3].

Figure 5 [4] shows local bearing failure which consists of a long zig-zag region between the yield point and the ultimate load. It can be implied that the hole is compressed and reformed after the compressed load, so it can absorb more load. Obviously, large hole deflection occurs. By contrast, the catastrophic failure graph in Figure 6 [4] has a short zig-zag region. After this zone, the acting load decreases rapidly because the broken plate cannot resist any load.

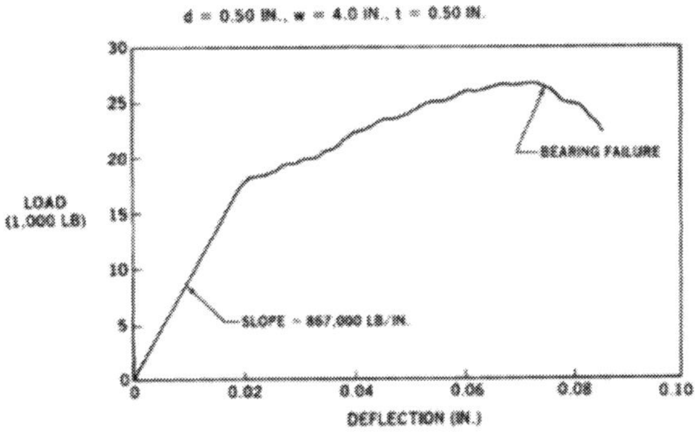

Figure 5. Non-catastrophic failure when the plate subjects to the bearing load (load-deflection graph) [4].

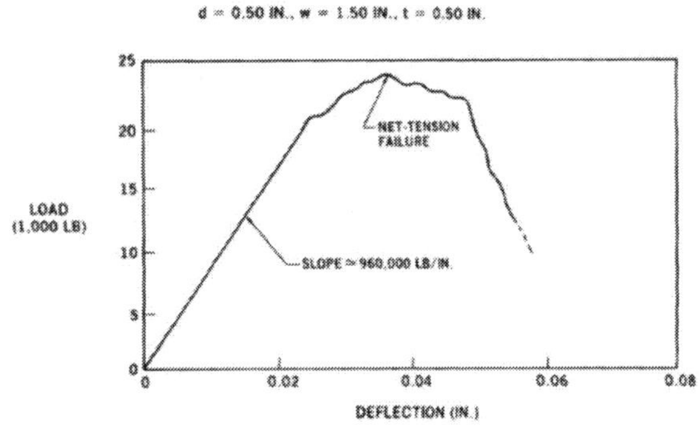

Figure 6. Catastrophic failure when the plate subjects to the bearing load (load-deflection graph) [4].

A number of researchers studying a composite plate with fastener holes usually define which mode is the most dominant. To avoid the catastrophic failure mode (including net tension and shear out which will occur when the plate is too narrow compared with the hole and the hole is too near the edge), many studies have attempted to define the safe size of the plate, especially the width and edge distance. Some researchers [5, 6] have claimed that the suitable plate width is around 2.5-4 times the hole diameter and the suitable distance from the edge to the hole center should be around 4 times the hole diameter. In addition, some research focused on analytical methods to define the failure modes. Characteristic curve method is a analytical theory which combines the hole failure modes with composite failure criterion has been used to estimate the maximum load and identify the failure modes in hole specimen [7].

Torsional resistance on bolt group is a theory to compute the resultant force occurring on the drilled hole from a vertical load acting on the other side of bolt group. It is calculated by moving the force from located on one end to the center of the bolt group. It leads to a coupled bending moment acting on its center, so two load types act to the drilled holes the removed-vertical force and the couple moment [8]. They are combined using the vector computation method. Figure 7 illustrates the method for the calculation of the force on the drilling holes.

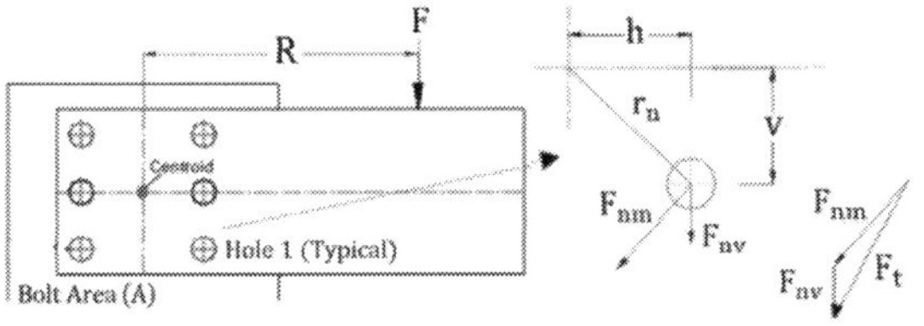

Figure 7. Diagram to calculate following the torsional resistance of bolted group [8].

A study [9] derived an equation for calculating the force acting on the double bolt using the torsional resistance bolt group theory. That paper varied the position of the drilling holes on the plate and calculated the force both from the vertical load and coupled moment. From this, the force can be plotted as a contour graph in which the X and Y distances shown by the vertical and horizontal axes, respectively, and the calculated force is the contour line. The force acting on the inner hole and the outer hole is illustrated in Figures 8 and 9. For example, when the X distance is 0.1 m and the Y distance is 0.01 m, the force action to the inner and outer holes is approximately 1.33 N and 2.32 N, respectively. It is evident that the resultant force on both holes gradually decreases when the distance between two holes is greater.

Due to the complexity and difficulty of generating the torsional resistance with the bending load, the experiment was simplified. The bending load causing the inclined compression load on the drilled holes was investigated using the bearing strength test following the ASTM standard on a hole drilled in Carbon Fiber Reinforced Polymer (CFRP) plates with the deviated ply orientation.

https://doi.org/10.1051/matecconf/201926401002

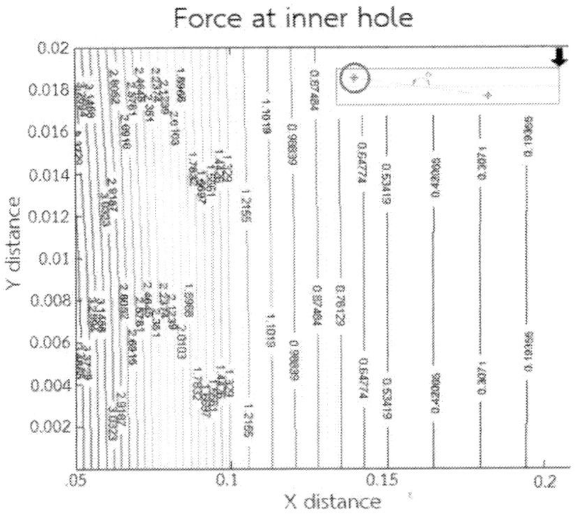

Figure 8. Force contour acting on the inner hole of the spar web with two fastening holes in offset misalignment [9].

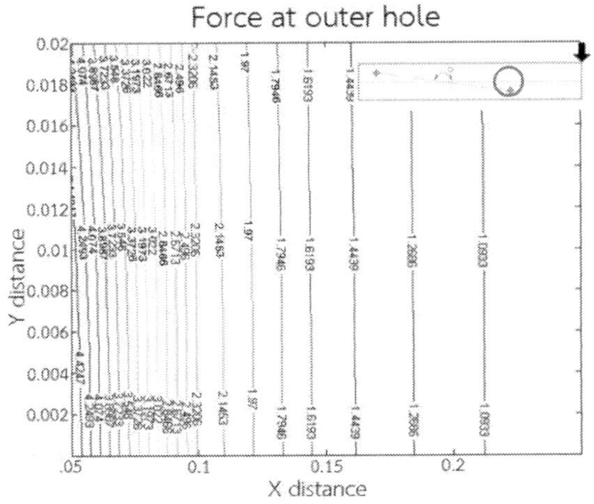

Figure 9. Force contour acting on outer hole of the spar web with two fastening holes in offset misalignment [9].

The fastener bolt hole experiment complied with the ASTM D5961 Procedure A standard [10]. The sizing of the specimen recommended in the standard is that the coupon should have a 0.25 in (6.35 mm diameter) hole with 19 mm distance from the nearest edge to the center and it should be cut to 140 mm long and 38 mm wide. Figure 10(a) provides an illustration of the recommended specimen sizing.

According to ASTM D5961 Procedure A, the testing specimen is clamped with a bolt to a double-lapped metal plate as shown in Figure 10(b).

Vangrimde and Boukhili [11] presented a method to measure the deformation of the drilling hole after it is drawn by the bolt. As can be seen in Figure 11, two Linear Variable Displacement Transformers (LVDT) are attached to the fixture and their probes touch the reference that pulls out the specimen. This procedure can eliminate the tensile plate displacement and focus on only the hole displacement.

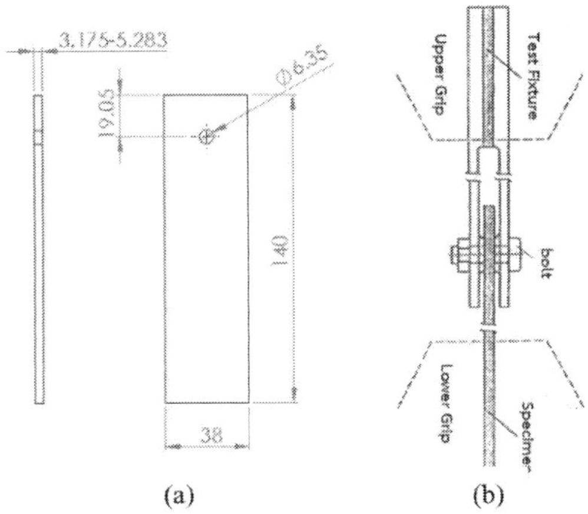

Figure 10. (a) Specimen dimension according to bearing test (ASTM D5961), (b) specimen with a bolt installation according to ASTM D5961 standard [10].

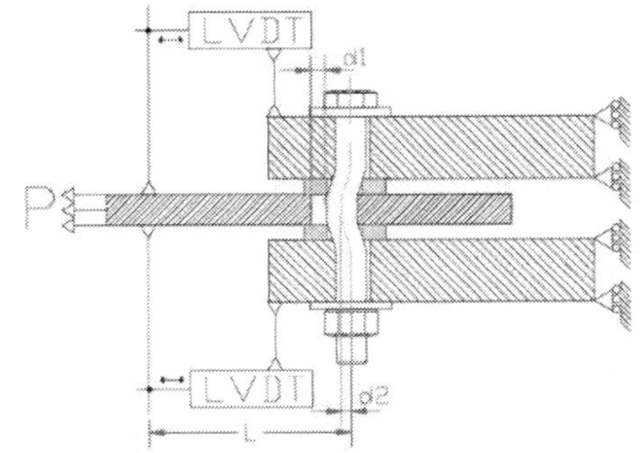

Figure 11. Vangrimde and Boukhili procedure to measure the hole displacement by the relation deformation of the specimen and the fixture using LVDT attached to the testing fixture [11].

Although the failure mode of a hole in the composite material has been studied in many fields, there is no research which focuses on the failure mode of composite plate fixing as a bolt group with the parallel offset misalignment concept. The acting load will deviate from the principle strength direction in the fabric. The current research investigated the bearing strength of the hole acting by using a simulated load of bending on the offset misalignment bolt group to a single force with the deviation angle on the hole. Consequently, the experiment was set up by testing carbon fiber-epoxy coupons with a different ply orientation according to the force deviation by the bending load following the ASTM standard.

2 Methodology

2.1 Specimen Lamination

According to ASTM D5961 Procedure A [10], the plates were fabricated using carbon fiber fabric twill 2x2 and EPOTEC YD 535 LV/TH 7253-8. Each lamina is 0.25 mm thick, so 14 layers of the coupon were laid in the same direction in order to obtain sufficient thickness for the bearing test following a vacuum infusion process under minus 1 bar-pressure. After fabrication, the plates were air cured for 24 hours. In order to provide greater accuracy on the degree orientation, which is a significant factor in this study, the plates were cut using CNC milling machine into 38 mm x 140 mm coupons with the different angles. Six sets of different angles were investigated (0, 5, 10, 15, 20, and 25 degree-deviation) as shown in Figures 12 and 13. The thickness of the coupons are in the range of 3.5±0.05 mm.

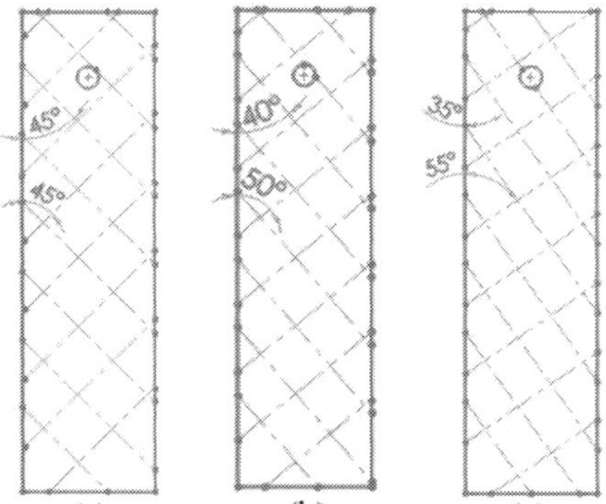

(a) (b) (c)

Figure 12. Drawing of samples after cured cutting with inclined angle (from ±45 degree) of (a) 0 degrees, (b) 5 degrees, and (c) 10 degrees.

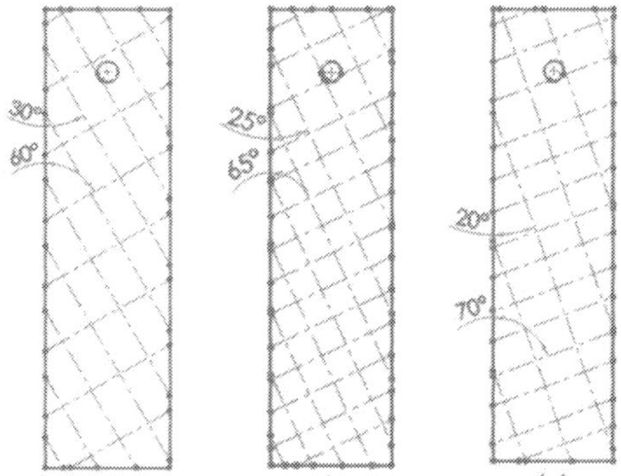

(a) (b) (c)

Figure 13. Drawing of samples after cured cutting with inclined angle (from ±45 degree) of (a) 15 degrees, (b) 20 degrees, and (c) 25 degrees.

2.2 Test Fixture

To comply with the ASTM standard mentioned above, a static tension load was applied to the coupon using a universal testing machine with tensile grips and a special fixture for the bearing test. The fixture was designed in the same manner as SACMA SRM 9-89 [12]. However, the standard gripping head of the testing machine is often too narrow (about 25 mm) to install the coupons following ASTM D5961. Therefore, the adaptor was modified to be able to clamp the 38-mm width coupon and to adjust the load alignment through the center of the machine head. The manufactured fixture and adaptor are illustrated in Figures 14 and 15.

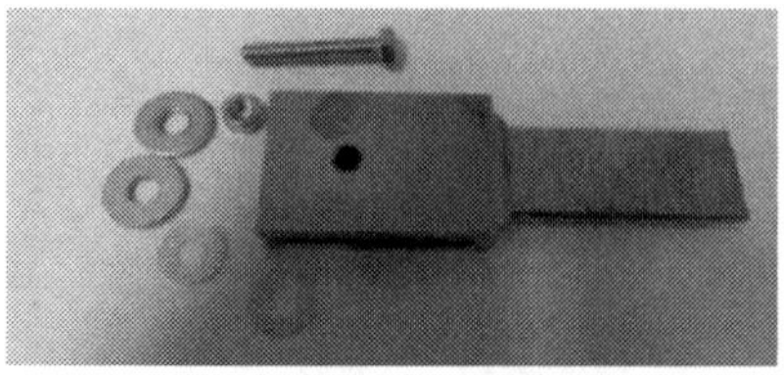

Figure 14. Specimen fixture with a bolted hole according to SACMA SRM 9-89 fixture.

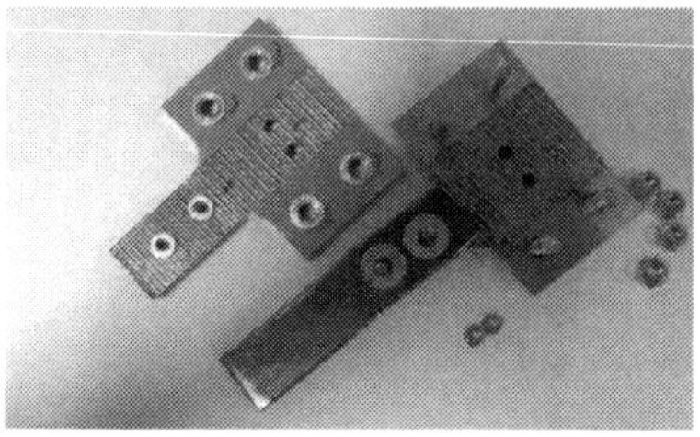

Figure 15. Adaptor for 38-mm width specimen and the 25-mm testing machine gripper.

2.3 Stress-strain measurement procedure

Due to coupling of the other extension portions in the coupon during the test following the Vangrimde and Boukhili procedure [11], a method to measure only the hole deformation is proposed. The extensometer is attached to the coupon in the zone next to the hole in order to measure the tensile extension of the coupon when the tensile load from the universal test machine is applied. As the gauge range is 25 mm, the total tensile deformation could be calculated by interpolation. Then, the hole deformation (δ_L) can be computed by deducting the tensile deformation (from the extensometer) from the total deformation (from the testing machine head). A diagram and all installation components including a sample are illustrated in Figure 16. The INSTRON universal testing machine and INSTRON extensometer 25 mm were used in this experiment. The load was applied in a displacement control protocol with 1 mm/min.

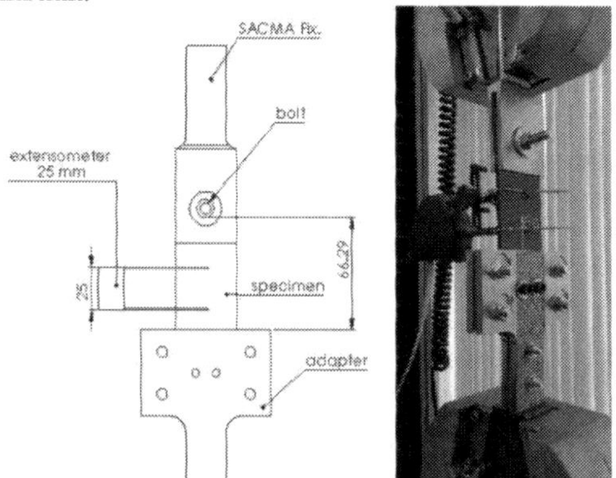

(a) (b)

Figure 16. (a) Testing component schematic including SACMA fixture, specimen with a hole, adaptor, and the extensometer and (b) the actual installation to the INSTRON universal testing machine.

https://doi.org/10.1051/matecconf/201926401002

2.4 Bearing stress-strain calculation

Load-displacement curves obtained from the specific test using a CFRP plate were recorded. According to ASTM D5961 [10], the bearing stress (σ_{br}) and bearing strain (ε_{br}) can be calculated using equation (1) and (2), respectively.

$$\sigma_{br} = \frac{P}{D \times t} \quad (1)$$

$$\varepsilon_{br} = \frac{\delta_L}{D} \quad (2)$$

When P = load reading from testing machine, N
D = bolt hole diameter, mm
t = plate thickness, mm
δL = hole deformation (by LVDT), mm

After the bearing stress-strain of each coupon was calculated, it was plotted to determine the bearing strength. The slope in the linear region of the stress-strain curve was defined as well as another linear curve, drawn by a 2% strain offset from the previous linear curve. The intersection of the bearing stress-strain curve from the experiment and the 2% offset line is called the offset strength which represents the yield strength of the hole. This point is known as the final stress limit of the structure. Figure 17 shows the bearing stress-strain and the method to determine the offset strength. Moreover, the slope of the plotted curve in the linear region is the bearing stiffness which is the mechanical properties of the laminate.

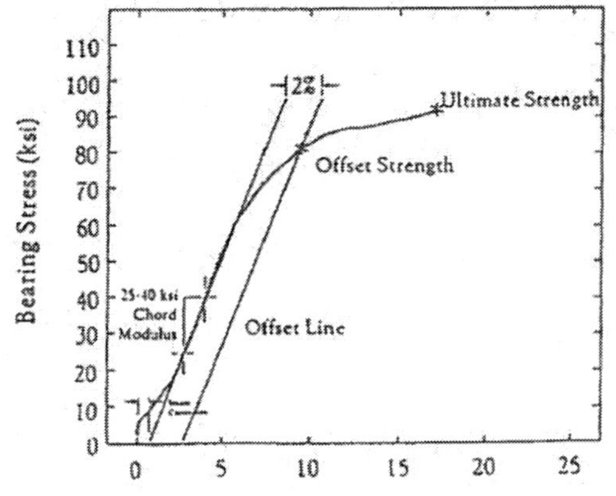

Figure 17. Bearing stress-strain test result analysis following ASTM D5961 to define offset strength, chord modulus, and ultimate strength [10].

3 Result and Discussion

3.1. Bearing stress-strain curves

After the load and displacement from the testing machine and extensometer were recorded, the bearing stress and bearing strain were computed by equation (1) and (2). There are two curves plotted on the graph. One has a larger strain which is calculated using the total strain measured with the testing machine. The other curve has the strain correction process that declines the strain. According to the elimination tensile extension approach, the corrected bearing strain was recalculated and plotted. Following this step, the 2% offset line was drawn on the corrected curve. The cutting point of the straight line and stress-strain curve is called offset strength. Figure 18 illustrated the example of data analysis. The darker line (0C total) is the raw data and the lighter line (0C adj) is the strain correction data. Then, the dashed line (Linear (offset)) is drawn to determine an intersection point between it and the strain collection curve in order to define the offset strength of the hole. It can be seen that the slope of the stress-strain curve increases because the tensile strain is eliminated, and the strain is computed from only the hole deformation. However, the stress level after strain correction remains the same.

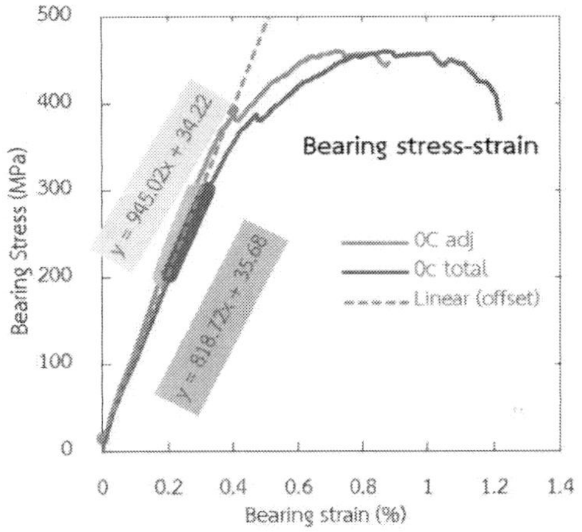

Figure 18. Bearing stress-strain curve obtain from the machine before (blue line) and after (orange line) bearing strain correction.

The stress-strain curves of three coupons for each deviation angle are shown in Figures 19 to 24. The three curves have a similar stiffness for the same deviation angle; the slope difference is less than 5% for each sample type. Thus, it can be implied that the bearing stiffness of each angle is similar. Because of the small deviation angles increase at 5 degrees per step, the bearing stiffness or slope of the curve for different deviation angles remains in a range (between 800 and 1100 MPa). Hence, the cutting point of every offset line should be at a similar stress level (around 350-450 MPa). Clearly, there are multiple peaks along each curve and every peak point refers to the maximum force that compresses the hole until it cannot bear and fail. As the strain increases, the CFRP material around the hole forms a new structure by the compression load and then it can absorb more load due to the natural inhomogeneity of the plate. It can bear the load until other damage failure (except bearing failure) occurs in the structure and it shows at the highest peak on the curves.

 https://doi.org/10.1051/matecconf/201926401002

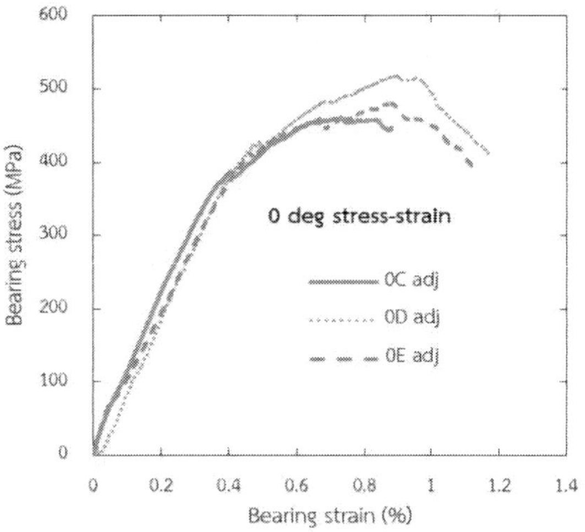

Figure 19. 0-degree inclination bearing stress/bearing strain test result after bearing strain correction.

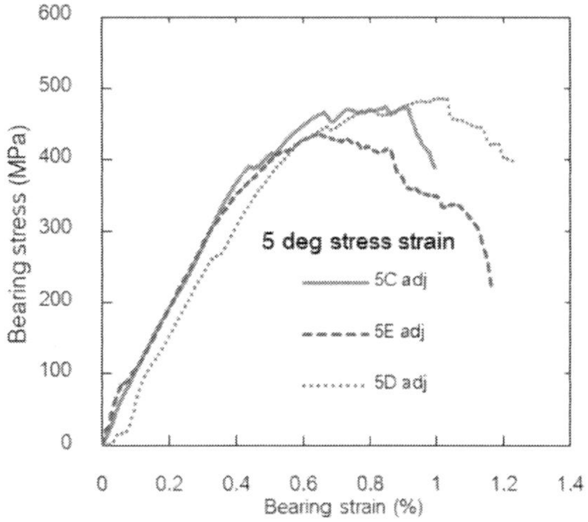

Figure 20. 5-degree inclination bearing stress/bearing strain test result after bearing strain correction.

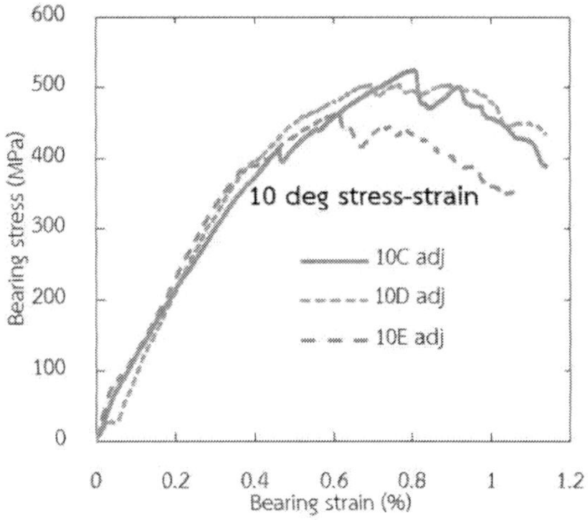

Figure 21. 10-degree inclination bearing stress/bearing strain test result after bearing strain correction.

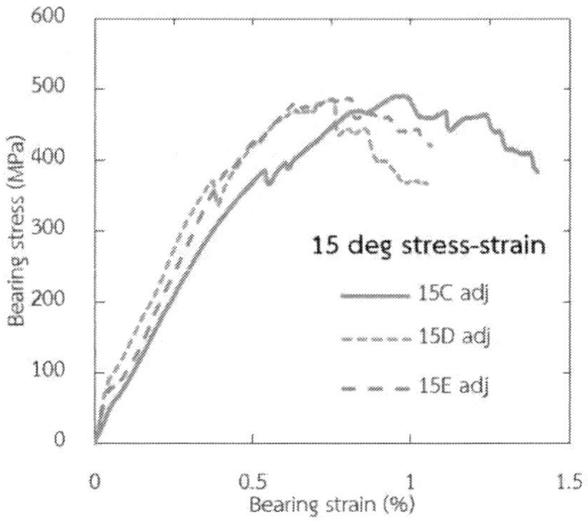

Figure 22. 15-degree inclination bearing stress/bearing strain test result after bearing strain correction.

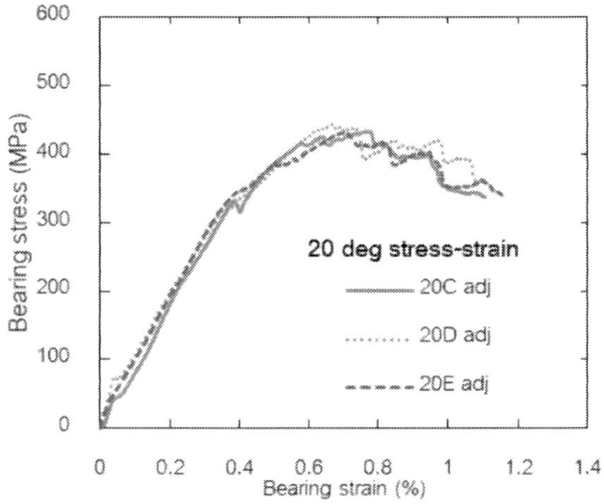

Figure 23. 20-degree inclination bearing stress/bearing strain test result after bearing strain correction.

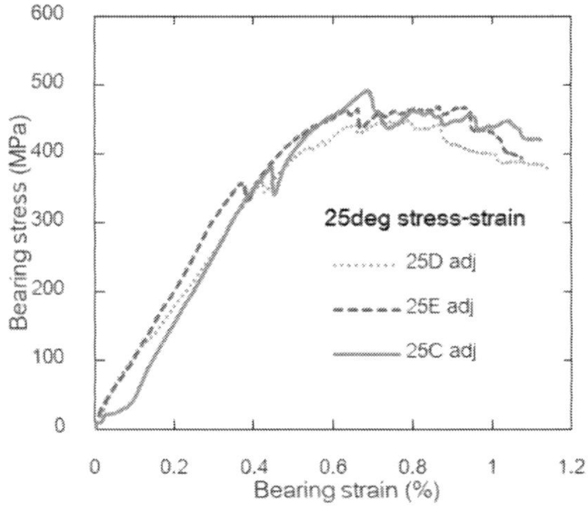

Figure 24. 25-degree inclination bearing stress/bearing strain test result after bearing strain correction.

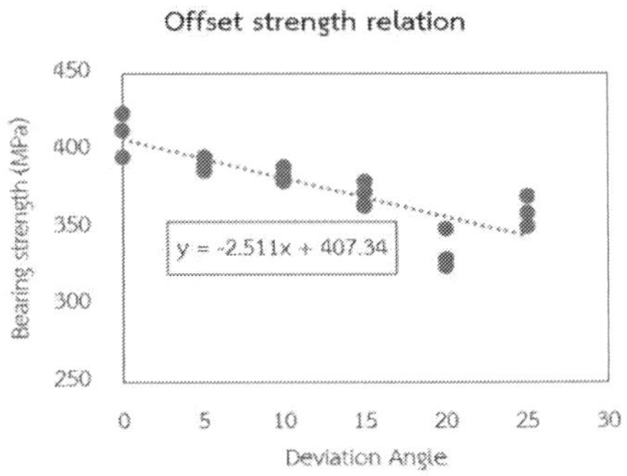

Figure 25. Offset strength result of each specimen and the linear regression equation.

3.2 Strength analysis

According to Figure 25, the offset strength seems to decrease when the deviation angle increases. Whereas the strength of 20-degree group remarkably declines. This may be because of a defect in the manufacturing process that caused this group to have a lower weight (approximately 1g) than other groups. The declination of strength can be interpreted by the resultant force in the fiber. As can be seen in Figure 26, the inequality between angles 1 and 2 results in some deviations in fiber direction. Thus, the bearing load in the vertical direction, which is distributed to both fibers, divides into two fibers unequally as can be seen in Table 1. The combination of force in two directions, which represents the capability to bear the amount of force on the structure, reveals that the resultant force of the larger deviation angle is lower than for the smaller deviation angle. This can be interpreted as the larger deviation angle results in a weaker structure because it diminishes the ability to absorb the total force acting on the hole. This is one of several reasons why the experimental results show the declination in bearing strength when there is a rise in the deviation angle.

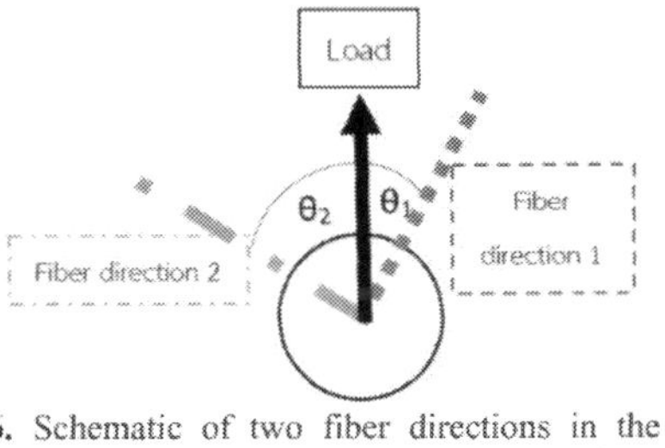

Figure 26. Schematic of two fiber directions in the woven fabric which deviated and subjected to the axial load.

Table 1. Inclination angle of fiber with the resultant force distributed to each direction.

Incline (deg)	θ_1 (deg)	θ_2 (deg)	$\cos\theta_1$	$\cos\theta_2$	Resultant (N)
0	45	45	0.707	0.707	1.41
5	40	50	0.766	0.642	1.42
10	35	55	0.819	0.573	1.39
15	30	60	0.866	0.5	1.36
20	25	65	0.906	0.422	1.32
25	20	70	0.939	0.342	1.281

3.3 Failure pattern

The comparison between 0-degree and 25-degree specimens showed a remarkable difference in hole failure. The hole deformation of 0-degree deviation (±45-degree ply orientation) was in the same direction as the acting force and the deformation gradually inclined with an increase in the deviation angle. The damage developed from the contact point between the bolt and specimen. This could then propagate along the fiber direction to the structural limit and lead to catastrophic failure. The assumption involved in this phenomenon is that the damage runs though the weaker direction which for the fiber is in the lateral direction. Figures 27 and 28 compare the specimens without and with inclination angle. The specimens without inclination angle can distribute force to both fibers equally resulting in the failure running straight along the force direction, but the specimens with an inclination ply angle (so that the fibers are aligned in the longitudinal direction) have greater strength, so the initial damage occurs in another direction. This phenomenon can be described by Figure 29.

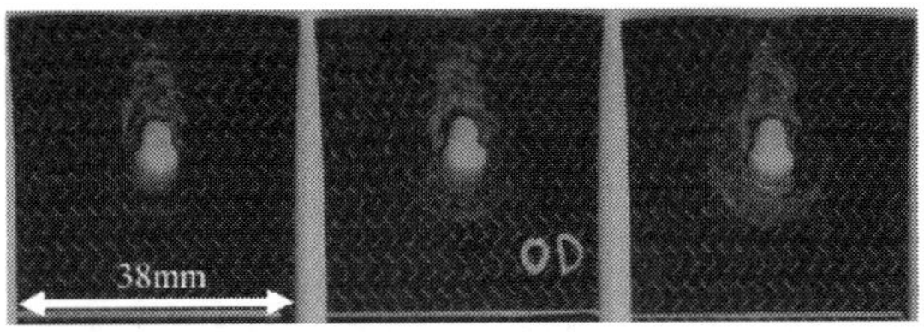

Figure 27. 0-degree deviation specimens' failure.

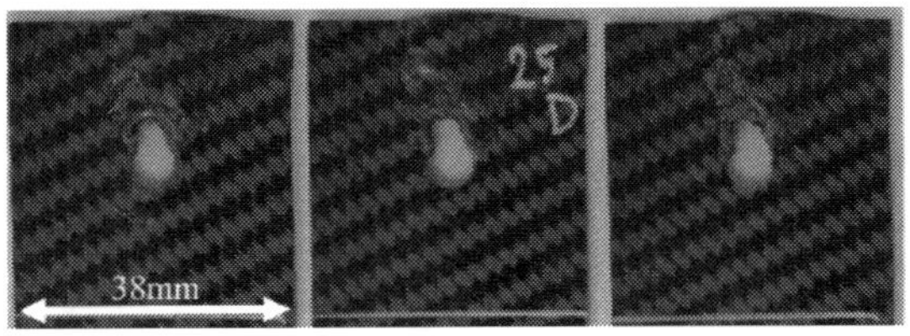

Figure 28. 25-degree deviation specimens' failure.

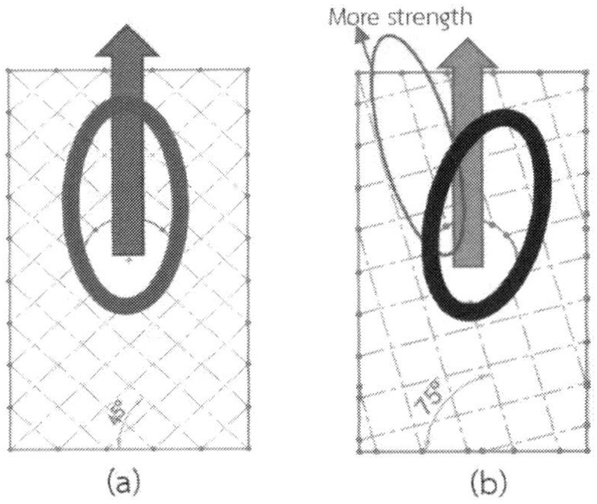

Figure 29. Schematic of failure mode with the fiber direction (a) 0-degree deviation angles and (b) 20-degree deviation angles.

https://doi.org/10.1051/matecconf/201926401002

3.4 Suitable drilling between holes

The wing spar fixture area is limited at the root zone. The possible drilling point was defined in a study [9] mentioned in the previous section. That paper presented a given resultant force on each hole's horizontal (X) and vertical (Y) coordinates which deviate from the lamination direction of the wing spar which normally consists of 45-degree lay-up woven fabrics. Following this, the inclined angle of the force acting on the drilling hole can be extracted from that study and the bearing strength of the CFRP in the deviation force direction can be computed from the linear relationship of the inclined angle and bearing strength in Figure 25. The most suitable points to drill two fastening holes are at the coordinates that give the highest the strength to the resultant force of the two holes. Eventually, the coordinate that gives the highest ratio is at the maximum horizontal position without vertical direction as shown in Figure 30.

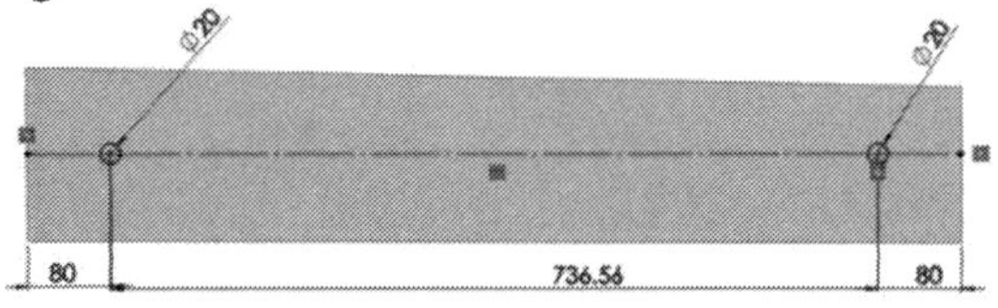

Figure 30. Suitable drilling hole positions after design considering the fiber inclination angle.

4 Conclusion

The bearing strength of the CFRP material when the force acts in the deviation angle from the setting orientation was determined by conducting an experiment following ASTM D5961. It was simplified by cutting the plate at an oblique angle instead of creating an inclination force. The results showed that the bearing strength of different deviation angles can linearly decrease when the inclination angle increases. This phenomenon can be explained by the force diagram that shows the deviation angle from cross-ply 45 degrees. The deviation angle reduces the ability to resist total force divided into each fiber direction which leads to a weakened structure. Moreover, the deviation angle makes the hole deformation deviate from the force direction because of the weakening in the direction that is not along the fiber axis. Eventually, a combination of the resultant force from the torsional resistance of bolt group equation and the linear relation between the inclined angle and bearing strength indicates the most suitable point to drill the fastening hole on the wing spar connection. The horizontal distance at the maximum point and at zero vertical distance is the most suitable location for this type of load.

References

1. Lancair International. *Center wing section installation,* URL: http://lancair.com/docs-lancair-legacy, accessed on 1/06/2017 (2017)
2. Falcomposite. *Furio In 27 RG*, URL: www.falcomposite.co/furio-rg.php, accessed on 1/06/2017 (2017)
3. C.C. Chamis, *Simplified procedures for designing composite bolted joints*. Journal of Reinforced Plastics and Composites, **9**, 614–626 (1990)
4. W.D. Nelson, B.L. Bunin and L.J. Hart-Smith, *Critical joints in large composite aircraft structure*. Technical report, McDonnell Douglas Corp (1983)
5. M.H. Ahmad, A.D. Crocombe and P.A. Smith, *Failure modelling of woven GFRP bolted joints under quasi-static loading*. International Journal Integrated Engineering, **5**, 2 (2013)
6. J. Ekh, *Multi-fastener single-lap joints in composite structures*. Ph.D. Thesis, Royal Institute of Technology, Sweden (2006)
7. F.K. Chang, R.A. Scott and G.S. Springer, *Strength of mechanically fastened composite joints*. Journal of Composite Material. **16**, 470–494 (1982)
8. Roymech, *Strength of bolts withstanding torsion generated shear loading*. URL: www.roymech.co.uk /Useful-Tables/Screws/Bolted-Joint.html, accessed on 1/06/2017 (2017)
9. N. Krajangsawasdi and P. Bunyawanichakul, *Analysis on optimal location of double-bolted joint with parallel misalignment subjected to eccentric load*. 55th Kasetsart annual conference, 436–442 (2017)
10. Departments and agencies of the department of defense, *MIL-HDBK-5H: Metallic materials and elements for aerospace vehicle structures* (1998)
11. B. Vangrimde and R. Boukhili, *Bearing stiffness of glass fiber-reinforced polyester: influence of coupon geometry and laminate properties*. Composite Structures, **58**, 57–73 (2002)
12. Wyoming Test Fixtures Inc, *SACMA bearing strength test (SACMA SRM 9-89). URL:* www.wyomingtestfixtures.com/Products/h2.html, accessed on 1/06/2017 (2017)

Structure integrated shape memory polymer composites for multi-dimensional forming

Tristan Timmel[1,a*], Alexander Hackert[2,b], Tomasz Osiecki[2,c], Colin Gerstenberger[2,d], Lothar Kroll[2,e]

[1]*Cluster of Excellence MERGE, Technische Universität Chemnitz, 09107 Chemnitz, Germany*

[2]*Institute of Lightweight Structures, Technische Universität Chemnitz, 09107 Chemnitz, Germany*

Abstract. Large-scale curved structures such as wind turbine wings usually require a special and cost intensive transport to the installation destination. These transport and installation costs can be reduced by a flat transport condition and the possibility of layering several structural components. For this reason, the focus at the Department of Lightweight Structures and Polymer Technology at TU Chemnitz was on a novel active material composite, which enables resource-efficient mass production and has a new component architecture. The large-volume multi-dimensional curvature of the active structure could be achieved by using a shape memory polymer (SMP). The associated reduction of the specific investment costs, the use of materials and the possibility of an integrative design, can contribute to the fact that, for example, the small wind turbines will become an economically viable investment in the future. The active structure influencing was represented by means of a finite element simulation (FEM) for different material composites and could be verified by generic demonstrators regarding its validity.

1 Introduction

Today most commonly used technological production of material compounds is the composite construction which combine self-sufficient materials, each with specific properties, in a construction using the right material in the right place. The materials are connected by a joining process using filler material. The mixed construction is thus characterized by a joining zone between the composite material systems of a composite material. This procedure can be realized especially by the differential construction. Due to the advantageous combination of different materials, the overall properties of an additively connected assembly should be improved. In this case, a material can dominate in its specific properties. The primary goal of using the mixed structure is to reduce the mass of an assembly while increasing performance. This increase is characterized by increasing the rigidity, the strength and the toughness and by improving the vibration resistance [1]. In addition, mixed construction solutions can be used to meet special customer requirements. In order to realize a cost-effective and mass-production-compatible structural element, an in-line production with continuous production is desirable. The goal is the development of an active composite material that can form a multi-dimensional structure with the help of an active matrix component. The multidimensional shaping is triggered by an external excitation, whereby the matrix component reacts with a change in length.

2 Active material compound

A particularly suitable matrix material is the shape memory polymer Desmopan® DP 2795A-SMP of the Bayer AG subsidiary Covestro Deutschland AG, which serve the plastics sector. This active component of the developed material composite controls the shaping behavior of the structure. The passive component is in the first step, a Polypropylene (PP) film which is replaced in the second step by a textile component, see Figure 1.

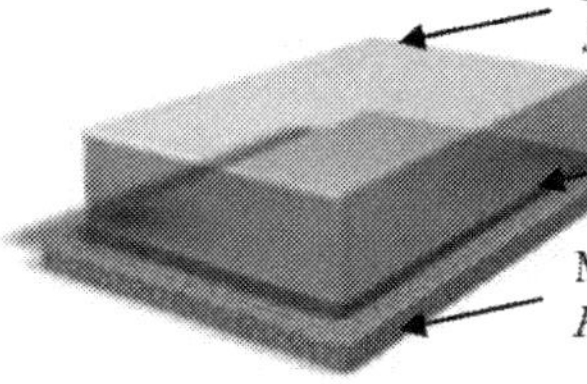

Figure 1. Illustration of the active material compound.

The textile material used for all variants is a pliable, impregnated plain weave cotton fabric, in which the warp and weft threads are made of twisted yarns. This textile fabric made of natural fiber with a surface weight of 320 g/m², as well as linen or hemp cloths, is used as sail cloth in the boat industry. The selected cotton fabric has a high tear strength, good weather resistance, resistance to UV radiation and good dimensional stability and ductility.

[a] Corresponding author: tristan.timmel@mb.tu-chemnitz.de

In addition, the textile is mildew resistant, dirt and water repellent. The decision to use the cotton fabric is based on the fact that this textile with the required properties is cheap and available in large quantities on the market. Figure 1 shows the two selected composite components, which are bonded by means of the joining process adhesive bonding to form the active material compound. This has the advantage that possibly occurring shear stresses in the interface can be reduced or compensated by the use of a suitable adhesive. The risk of delamination of the composite material can thus be minimized and the composite properties are controlled.

3 Structure formation

3.1. Analytical consideration

The basis for the modeling of the generic technology demonstrators is the mathematical description of the active material composite. According to the functional diagram shown in Figure 2, the composite can be transformed into a model of a contracting actuator and a linear compression spring.

a) Block diagram b) Schematic illustration

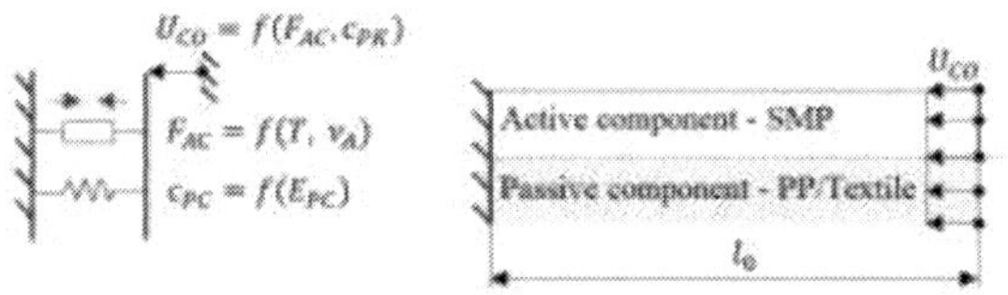

Figure 2. Functional diagram of active material combination.

Starting from the equilibrium condition that the contraction force of the active component F_{AK} of the force of the passive component F_{PK}, the contraction shift of the active composite U_{CO} can be described below

$$U_{CO} = F_{PC}/c_{PC} = (l_0 \cdot A_{AC} \cdot (K_{\sigma 1} \cdot T - K_{\sigma 2})) / (E_{PC} \cdot A_{PC}). \quad (1)$$

The desired strain (contraction) of the active composite S_{CO} is given in the following:

$$S_{CO} = U_{CO}/l_0 = (A_{AC} \cdot (K_{\sigma 1} \cdot T - K_{\sigma 2})) / (E_{PC} \cdot A_{PC}). \quad (2)$$

The resulting deflection due to the induced structure curvature is subsequently derived analytically. The analytical solution is based on the thermal expansion problem in bimetallic beams, see Figure 3. The radius of the active component R_{AC} is proportional to the radius of the second component R_{PC} and equal in the region of the joint zone. The two radii are summarized approximately to a radius of curvature R_C

$$R_{AC} \sim R_{PC} = R_C. \quad (3)$$

With the approach of Eq. (3), the same curvatures κ are subsequently produced in the actuator layer and the second component.

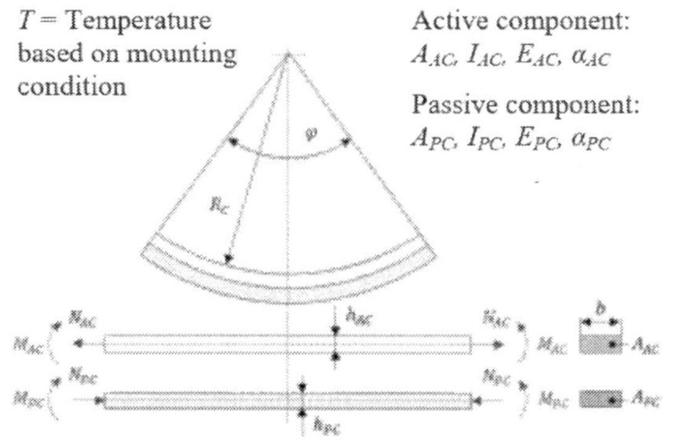

Figure 3. Thermal bending model.

The product of the modulus of elasticity E of the respective component and the area moment of inertia I gives the bending stiffness, which in both layers is defined as the ratio of bending moment to curvature M/κ

$$\kappa = 1 / R_C = M_{AC} / (I_{AC} \cdot E_{AC}) = M_{PC} / (I_{PC} \cdot E_{PC}) \quad (4)$$

The bond is defined by a longitudinal force equilibrium at the structural ends with the longitudinal force of the active component N_{AC} and the longitudinal force of the passive component N_{PC}. The sum of the bending moment of the active component M_{AC} and the passive component M_{PK} gives the total bending moment of the composite M

$$N_{AC} = N_{PC} = N \rightarrow N \cdot (h_{AC} \cdot h_{PC}) / 2 = M_{AC} + M_{PC} = M \quad (5)$$

The displacement compatibility at the interface with the thermal expansion coefficients α of the respective component as well as the cross-sectional areas A_{AC} and A_{PC}

$$\alpha_{AC} \cdot \Delta T + N / (E_{AC} \cdot A_{AC}) + h_{AC}/2 \cdot 1/R_C \quad (6)$$

$$\alpha_{PC} \cdot \Delta T - N / (E_{PC} \cdot A_{PC}) - h_{PC}/2 \cdot 1/R_C \quad (7)$$

can be formulated with the moment of inertia of the active component I_{AC} for the desired structure curvature.

After replacing the thermal expansion difference by the contraction of the active component from Eq. (2) taking into account the thermal expansion of the second component, the resulting actuation strain A is given in Eq. (8)

$$A = \alpha_{PC} \cdot \Delta T + S_{CO} f(T)$$

$$A = \alpha_{PC} \cdot (T - R_T) - (A_{AC} \cdot (K_{\sigma 1} \cdot T - K_{\sigma 2})) / (E_{PC} \cdot A_{PC}) \quad (8)$$

and the sought-after active structural behavior in Eq. (9)

$$1 / R_C = (\alpha_{PC} \cdot (T - R_T) - A_{AC} / (E_{PC} \cdot A_{PC}) \cdot (K_{\sigma 1} \cdot T - K_{\sigma 2}))$$
$$/ [2 / (h_{AC} + h_{PC}) \cdot I_{AC} / A_{AC} \cdot (1 + (I_{PC} \cdot E_{PC}) / (I_{AC} \cdot E_{AC}))$$
$$\cdot (1 + (A_{AC} \cdot E_{AC}) / (A_{PC} \cdot E_{PC})) + (h_{AC} + h_{PC}) / 2]. \quad (9)$$

https://doi.org/10.1051/matecconf/201926401003

The deflection U_{bend} of the active cantilever is shown schematically in Figure 4 and can be determined with the help of Eq. (10).

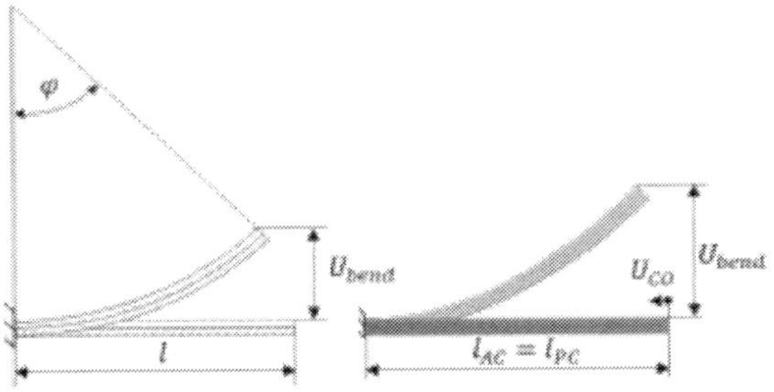

Figure 4. Bending model.

The deflection U_{bend} is defined using the angle of curvature φ below:

$$U_{bend} = d\varphi / dx \cdot l\text{^}2 / 2$$

$$d\varphi / dx = const. = 1 / R_C \rightarrow 1/R_C \cdot l\text{^}2 / 2 . \quad (10)$$

In summary, it can be stated that the structural properties of the second component have a major influence on the structure curvature of the composite material due to a change in length of the active component.

3.2. Numerical consideration

The material composite demonstrator, which is to be used for the simulation of structural shaping, consists of two components, which are connected to each other by means of an adhesive. The active component consists of the SMP Desmopan DP 2795A-SMP and has the task of producing a deflection of the material composite in the Z direction by a controlled change in length.

The second component represents the passive cover layer of the composite material and consists of a textile fabric. The geometry and dimensions of the composite material demonstrator result from the SMP tensile specimens according to DIN EN ISO 27, type 1A. The desired structure curvature of the composite material, as a result of a change in length of the active component, requires contacting of the two components in at least two points. With increasing contact, up to the full-surface connection, an increase in the degree of deformation and a more uniform curvature is expected.

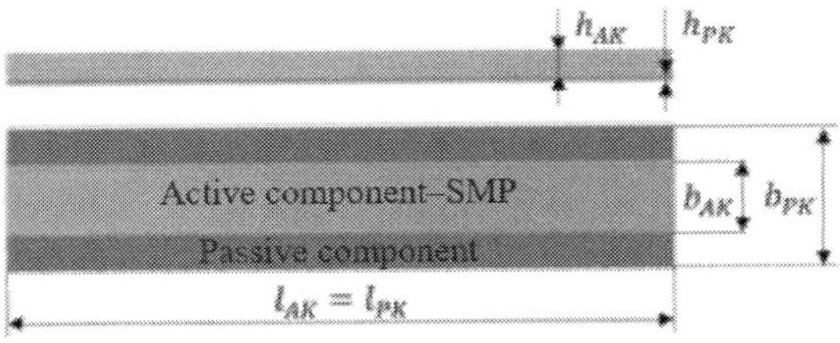

Figure 5. Dimensions of the generic technology demonstrator

The active component of the generic composite material demonstrator is first heated to a temperature above the switching temperature T_g in order to solve the physical crosslinking (secondary valence bonds). In this thermally activated state, the SMP is elastically deformable [2]. The Desmopan DP 2795A-SMP tensile bars are heated in the climate chamber of the Zwick / Roell Z250 static double-axis tensile testing machine from Zwick / Roell and stretched by 2.85 mm in a strain-controlled manner, see Step 2 in Figure 6. By means of the multiXtens tactile extensometer, exact and reproducible stretching is possible. By cooling the SMP below the switching temperature T_g, the temporary shape of the tensile specimen can be frozen by the formation of physical crosslinks and the concomitant increase in rigidity in the hard segment of the block copolymer [3]. Subsequently, the tensile specimen is cut to the specified length dimension l_{AC} and joined to form a composite material.

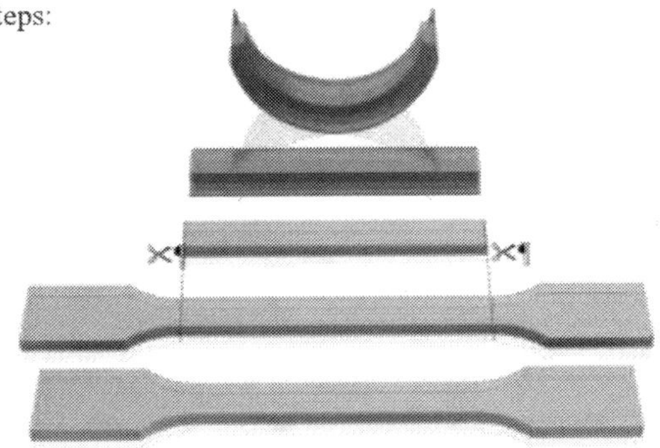

Figure 6. Creation of the generic technology demonstrator

Subsequently, when the SMP is reheated to a temperature greater than T_g, the physical crosslinks dissolve again, returning the tensile sample to its permanent shape. Step 5 in Figure 6 symbolically shows the resulting deflection due to the contraction of the SMP by 2.85 mm for a composite material variant, which is evaluated in the next step. In the following composite material variant in Figure 7, the SMP component is bonded to the entire surface of the 0.5 mm thick polypropylene film (PP). Due to the predetermined temperature change of 30 K, the simulated contraction of the SMP is 2.85 mm and leads to a deflection in the Z direction of 41 mm.

Numerical results: z-displacement in [mm]

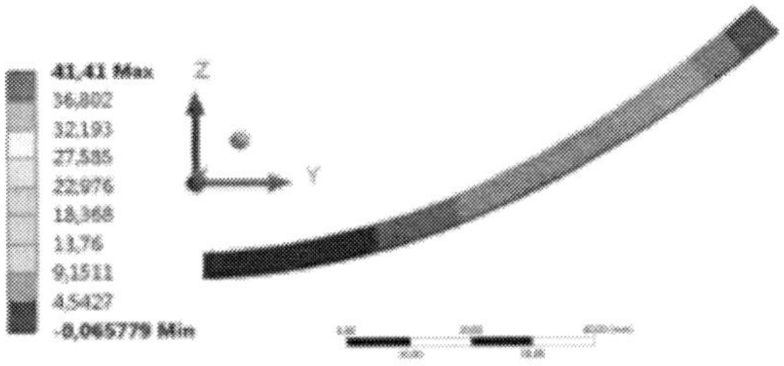

Figure 7. SMP-glue-PP – Full-surface contact.

https://doi.org/10.1051/matecconf/201926401003

The analytical consideration has already shown that when using a nearly limp component an induced structure curvature of the composite material due to a change in length of the active component is theoretically not possible. Despite the use of a nearly limp component, a structural formation on the generic demonstrators can be watch. This is due to the adhesive used. The glue impregnates the fibers of the textile semifinished product in the region of the joining zone, which means that the fibers are protected against buckling under a compressive load (compression). This leads locally to an increase in the E-module in this composite material. The adhesive thus assumes the function of the matrix in a composite material or a fiber-reinforced plastic composite. A further increase in the modulus of elasticity is accompanied by the increase in the strength of the adhesive due to the chemical reaction in the curing process. The resulting Z-displacement of the textile material composite with a full-surface adhesive layer between the two components is shown in Figure 8.

Numerical results: z-displacement in [mm]

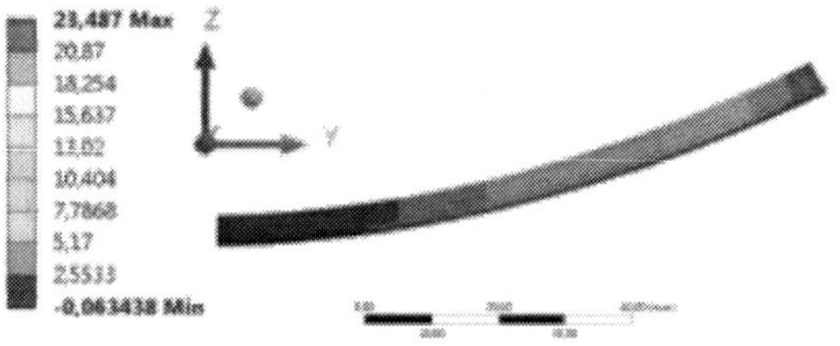

Figure 8. SMP-glue-textile – Full-surface contact.

3.3. Experimental consideration

The evaluation of the experimentally determined Z-displacement for the generic composite material with PP film and textile fabric as cover layer is shown in Figure 9 and also in Figure 10 as a box plot. In the experimental studies the generic demonstrators with full-surface contact showed a large structural formation in the Z direction.

Experimental results: z-displacement in [mm]

SMP-glue-PP: U_{bend} = 39 mm

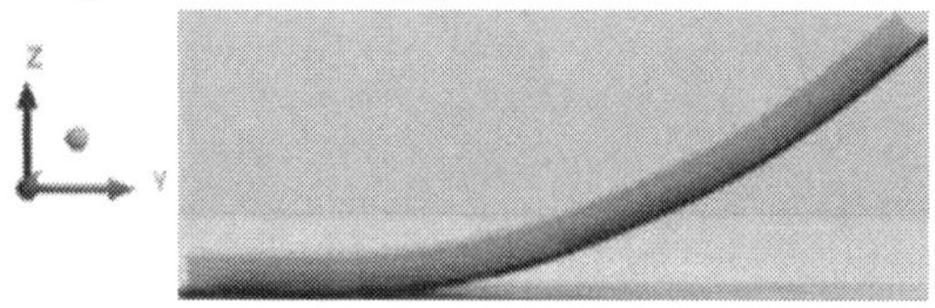

SMP-glue-textile: U_{bend} = 25 mm

Figure 9. SMP-Glue-Textile – Full-surface contact.

These two composite material variants are compared with respect to the Z-displacement.

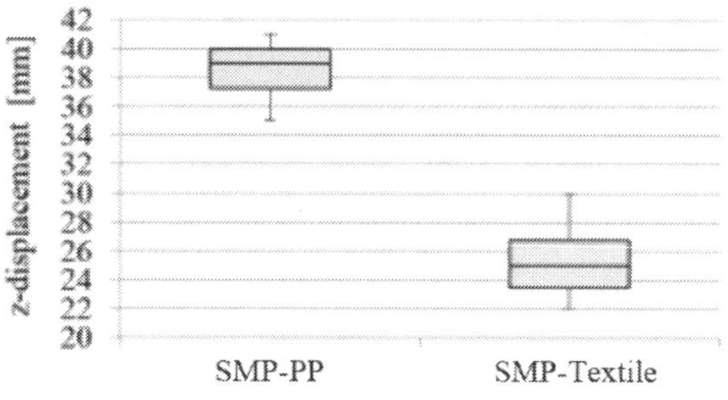

Figure 10. Evaluation of the collected z-displacement.

The deviation of the Z-displacement in the range of 22 mm to 30 mm is due to the different degree of impregnation of the textile with adhesive and the associated layer thickness of the adhesive. The Z-shift of the textile material composite is on average 35% lower than that of the isotropic material composite with PP film. The reason for this lies in the textile properties and can be adjusted by, for example, the type of fabric, thread and stitch density. The validity of the finite element model could also be proven for the textile application. For initial estimates of the structural behavior of active material composites, the FEM model used is sufficiently accurate and thus forms the basis for future investigations.

4 Summary

The use of a textile wind turbine blade with active structural curvature shown in Figure 11 is one way that can be used in the near future. The advantages that speak for the use of a textile wind turbine blade, in addition to the low weight and ease of manufacture, the possibility of functional integration and personalization. The functional integration can be implemented easily and inexpensively by means of embroidery technology and adhesive technology. The back injection technology is also a possibility to integrate sensors or actuators with a bio-based thermoplastic material [4] on the textile component.

Figure 11. Small wind turbine with SMP composite blades.

Sensors such as embroidered strain gauges, temperature sensors or moisture-detecting sensors can be applied by means of embroidery on the textile. The SWT can thus simultaneously function as a weather station. Design features and personalization options such as actively illuminating fibers or Q-LED lighting elements are currently the subject of research at the Chemnitz University of Technology and can also be applied to the textile sail for future applications. The use of flexible photovoltaic cells is also conceivable on the textile wind turbine wing. Due to the vertical clamping of the sails, the diameter of the rotor is reduced, whereby the double-clamped sword segments are subjected to a bending and thus cause a buckling of the individual sails [5]. The applied SMP profiles on the canvas support the stretched-out shape and help to prevent the sail breaking through in headwinds. In addition to the sail shape stability, the profiles have the task to guide the incident wind at an optimum angle to the sail area. By horizontally directed laminar flow and forwarding this in the wind-averted wing, the efficiency of the rotor can be significantly improved. The fundamental advantage of the developed textile shape memory polymer wind turbine blade is the even transport state, which allows stratification of many wind turbine blades and thereby et al reduced transport and installation costs. The large-volume curvature of the wing thus takes place only in the assembled state at the site. The envisaged integration of the wind turbine in an already electrified and developed light mast reduces u. a. the costs for the foundation, the development and commissioning of the small wind turbine. In the next steps of this research project the goal is to pursue the strategy of bivalent resource efficiency (BRE strategy) more closely in order to exploit all available resource efficient.

ACKNOWLEDGEMENTS

The presented studies were performed as part of a pre-investigation within the Federal Cluster of Excellence EXC 1075 “MERGE Technologies for Multifunctional Lightweight Structures“ and supported by the German Research Foundation (DFG) to explore integrative in-Mold processes with plastic/ metal composites. Financial support is gratefully acknowledged.

References

1. T. Timmel, A. Hackert, T. Osiecki, C. Gerstenberger, L. Kroll, *Combined Injection Molding Technology for Dynamically Stressed Multi-Material Coupling Elements*, Key Engineering Materials, **vol. 744**, pp. 322-326, (2017)
2. M. Behl, A. Lendlein, *Shape Memory Polymers*, Materials Today 10, **vol. 4**, pp. 20-28, (2007)
3. C, Schmidt, *Werkstoffwissenschaftliche Untersuchungen zur Verarbeitung von Formgedächtnis-Polymeren und zur funktionellen Ermüdung bei zyklischer Abfrage des Einwegeffekts*, PhD, (2011)
4. R. Rinberg, T. Hartmann, A. Nikiforov, L. Volfson, L. Kroll, *Investigation of bio-based polyamide with short fibers for lightweight structures*, Conference proceedings International MERGE Technologies Conference - IMTC 2017 Lightweight Structures, **vol. 3**, pp. 33-34, (2017)
5. L. Kroll, M. Heinrich, S. Müller, J. Kaufmann, *FE-Analyse anisotroper Koppeleffekte mit der ESAComp-Schnittstelle zur belastungsgerechten Auslegung von Snowboards*, ANSYS Conference & 25. CADFEM Users´Meeting, **vol. 25**, (2007)

Flaking of PEEK under one-point rolling contact fatigue using Al_2O_3 ball

Hitonobu KOIKE[1a], Genya YAMAGUCHI [2], Koshiro MIZOBE[2], and Katsuyuki KIDA[2]
[1] *University of Miyazaki, 1-1 Gakuen Kibanadai-nishi, Miyazaki, 889-2192, Japan*
[2] *University of Toyama Gofuku 3190, Toyama, 930-8555, Japan*

Abstract. The growth of flaking as tribological fatigue failure in PEEK was investigated through the one-point type rolling contact fatigue test between a machined PEEK polymer shaft and an alumina bearing's ball. Due to Hertzian contact of cyclic compressive stress, the subsurface fatigue cracks in the PEEK shaft propagated in rolling and axial directions. When the rolling fatigue life of the PEEK shaft reached 10^6 fatigue cycles, many narrow angled cracks occurred in the near-surface of the rolling track without flaking. On the other hand, when the flaking ocuurred on the PEEK shaft before 10^6 fatigue cycles, semicircular surface and subsurface crack propagations were observed. From these observations, it was found that micro-flaking occurred due to the linkages between subsurface –and surface cracks. Flakingdeveloped due to the accumulation of these micro-flakings.

1 Introduction

Lightweight function parts using premium plastics are extended in various mechanical elements under special chemical environments [1-2]. Polymer bearings or gears as precision-machined custom functional parts are suitable for assembly robots, agricultural devices, applications related to chemical and medical food [3]. Friedrich et al. [4] stated that the most important points of the polymer component design are mechanical stresses and environments.

Poly-ether-ether-ketone (PEEK) has high chemical resistance, self-lubrication ability, no degradation risk in salt water, and non-electrical conductivity. Researchers investigated tribological failures of PEEK materials for frictional parts used under rolling contact fatigue (RCF) [3,5-7]. Flaking as tribological fatigue failure causes noise, vibration, or alignment error in components, which is an important quality problem for products. Marshek examined failure in plastic bearings, and explained flaking as material breaking loose from the surface in the form of shallow flakes [5]. Briscoe stated that a fatigue process with repeated stress cycles could involve crack propagation. [6]. Avanzini et al. reported that in RCF between a PEEK roller and a steel disk micro-pitting and transversal crack on the rolling surface of the roller edge occurred due to tangential stress [7]. Dearn et al. reported that the location of the maximum shear stress depended on the friction condition and was related with the fracture of PEEK polymer gears [8]. Flaking was related to the internal fatigue cracks which were caused by cyclic shear stress below the contact surface [9]. Koike et al. stated that in the RCF of the PEEK-alumina ball bearings with an artificial pit, the subsurface fatigue crack ocuured at 100-200 um depth from the contact surface. This means that the internal fatigue crack from an artificial pit was influenced by internal shear stress [9]. As Young's moduli of polymers are significantly lower than that of common metals or ceramics, internal stress and elastic deformation of friction area of polymers are greatly influenced by cyclic Hertzian contact. Therefore, the behavior of the internal fatigue crack propagation in the machined PEEK elements is not well understood. The practical research on PEEK's tribological fatigue fracture in application of PEEK and ceramic parts is investigated in few studies [9-11].

Kida et al. performed RCF test using Si_3N_4 balls and Si_3N_4 plates with an artificial pit, and investigated the fatigue crack propagation under the rolling track [12]. They proposed that the subsurface crack propagation depended on the mode II or mode III crack propagation after the surface crack propagation in sliding-rolling contact [12].

In this study, in order to explore the flaking mechanism under rolling contact between machined PEEK shaft and alumina ball, one-point Rolling Contact Fatigue (RCF) tests were performed. The internal fatigue crack propagation in early stages of the flaking was observed.

2 Experimental procedure

2.1 PEEK shaft specimens

The machined PEEK shaft with a 290 mm length and a 17 mm diameter was used for the one-point RCF test (Fig.

[a] Corresponding author: koikeh@cc.miyazaki-u.ac.jp

1). The PEEK material grade was 450G in VICTREX™. An alumina ball with 9.525 mm (3/8 inch) diameter was used as a counter contact part in the one-point RCF test. To estimate the contact stress between the shaft and the ball, Young moduli of PEEK (3.5 GPa) and alumina (360 GPa) were used. After the grinding process on the PEEK shafts, the average roughness (R_a) on the surface of the shaft was 0.69∼1.10 μm.

2.2 One-point rolling contact fatigue test

The RCF tests were performed by using the one-point radial loading machine as shown in Figure 1 [13]. The alumina ball contacted the PEEK shaft specimen under oil lubricant condition. Kinematic viscosity of the lubricant oil was 46 mm^2/sec at 40 °C. Radial load was applied to the PEEK shaft, while rotating by a DC motor. The rolling track was concentrated on only one line on the PEEK shaft under a radial load. This means that we can observe easily the damage zone of the parts under cyclic contact stress. In the one-point RCF tests, the contact area between the PEEK shaft and the alumina ball was ellipse shape due to Hertzian elasticity contact (detail A in Fig. 1b). The maximum Hertzian contact stress P_{max} values and contact ellipse diameter values were shown in Table 1. The test conditions of radial load and rotation speed were 90-360 N and 1200 rpm, respectively. They were the same as our previous works [9,14]. This is because the present tests were done in order to focus on the details of the fatigue fracture origin/propagation of the one-point RCF test. The surface velocity of the PEEK shaft was estimated as 1068 mm/sec. The minimum film thickness (h_{min}) in the point contact at 120 N was caluculated by Hamrock and Dowson's equations [15]. The h_{min} was 0.142 μm, which corresponded to the mix lubrication. The one-point RCF tests were continued until 4.5×10^6 fatigue cycles or fatigue failure. The flaking growth and fatigue crack propagation in the tested PEEK shaft specimens were observed by laser confocal microscopy (Fig. 2a and 2b). The PEEK shafts after the RCF tests were cut along the rolling contact track. The cross-sections of the rolling track were polished, and the growth of flaking was arranged by using 2.5D layer observation method (Fig. 2c).

3 Experimental results and discussion

3.1 Contact surface of PEEK shaft under one-point RCF

Figure 3 shows microscope images of the rolling track of the PEEK shaft after the one-point RCF tests. The higher the radial load, the larger the flaking size. As shown in Fig. 3(b), semicircular surface cracks like 'fishscale shape' were observed near the flaking. The rolling direction length (L_{ro}) of flaking was longer than the axial direction width (W_{ax}). This means that in flaking on the rolling track of the PEEK shaft the fatigue cracks propagated faster toward the rolling direction than toward the axial direction. When radial load was 360 N, the delamination occurred on whole rolling track (Fig. 3(d)). Therefore, L_{ro} and W_{ax} could not be measured. It was thought that subsurface cracks propagated due to internal shear stress. Table 2 shows the

Table 1. Maximum contact pressure and contact ellipse diameter values.

Radial applied load, Q [N]	Maximum contact pressure, P_{max} [MPa]	Long diameter, D_L [mm]	Short diameter, D_S [mm]
90	274	0.918	0.684
120	301	1.011	0.753
240	380	1.273	0.949

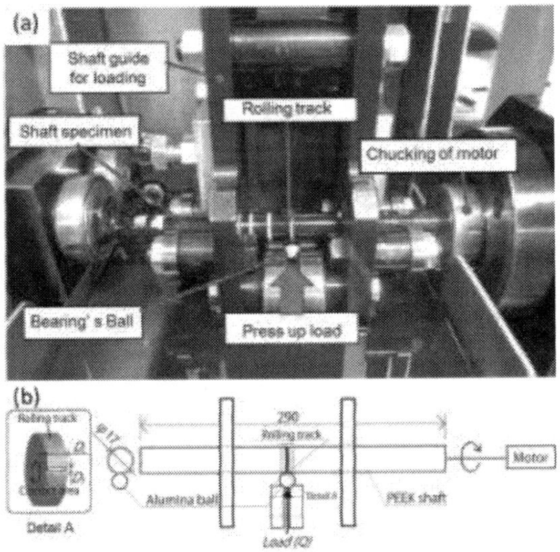

Fig. 1. One-point radial loading test machine; (a) photo image of one-point RCF test, (b) illustration of test specimen.

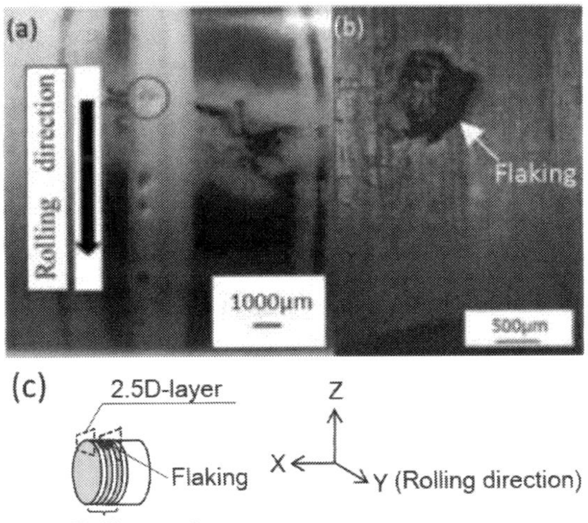

Fig. 2. Test specimen after one-point RCF test; (a) flaking on the rolling track, (b) close-up of flaking, (c) illustration of 2.5D layer observation (in Fig. 5).

 https://doi.org/10.1051/matecconf/201926401004

flaking size at 1200 rpm. This result suggests that the higher the radial load, the shorter the fatigue life. The flaking at 240 N load occurred in early operation term compared to at 90 N load because the crack propagation at high applied load was faster than that at low load.

3.2 Crosssection of PEEK shaft under one-point RCF

Figure 4 shows 2.5D layer images of the crosssection of the flaking area along rolling direction at 1.9×10^6 fatigue cycles under a load of 240 N. The maximum depth of the flaking was approximately 215 μm, and the downward branch subsurface crack propagated along the rolling direction (Fig. 4(#1)). The flaking bottom gradually rose upward along axial direction (Fig. 4(#1)~(#4)). Additionally, the downward branch subsurface crack from the flaking bottom was observed at 268 μm depth as shown in Fig. 4(#5). This result shows that the downward branch subsurface crack influenced the growth of flaking.

Figure5 shows 2.5D layer images of the crosssection of the flaking at 2.9×10^4 fatigue cycles under 240 N radial load. As shown in Fig. 5(#2), downward branch subsurface crack was observed below the 2nd flaking. By the propagation of subsurface cracks, the first flaking merged with the second (Fig. 5(#2 and #3)). At this time the maximum depth of the flaking was 534 μm. As shown in figure 5(#8), the flaking grew due to the fusion of subsurface crack and surface crack.

Figure 6 shows the close-up of the flaking origin in the nearsurface of the PEEK shaft. In early stage of flaking, micro-pitting or micro-flaking occurred in the nearsurface of the rolling track. The first micro-flaking occurred within 200 μm depth from the rolling track surface due to the fusion of some subsurface cracks (Fig. 6b). It was found that flaking grew gradually with the linkages of the micro-flakings.

Figure 7 shows the profiles of the maximum depths of the flakings at 120 N. Test #2 corresponds to the flaking in Figure 4. In the profiles, valley points were observed at flaking bottoms. This means that the flaking gradually grows due to the accumulation of small (micro) flakings.

Figure 8 shows the microscope image of the cross-section around non-flaking area on the rolling track of the PEEK shafts at 1.9×10^6 cycles under 120 N. Many narrow angled cracks were observed on the subsurface. The narrow angled cracks propagate along the rolling direction at around 200 μm depth from the shaft surface. These cracks caused micro-flaking.

Table 2. Flaking size of the PEEK specimens.

Test number	Radial load Q [N]	Rolling direction length, L_{ro} [mm]	Axial direction width, W_{ax} [mm]	Maximum depth of flaking [mm]	Fatigue cycles N [cycles]
1	90	0.759	0.450	0.154	4.5×10^6
2	120	0.866	1.077	0.215	1.9×10^6
3	120	1.465	1.227	0.413	4.6×10^5
4	120	1.510	0.930	0.610	3.6×10^5
5 [ref. 11]	120	2.039	1.416	0.682	4.1×10^5
6	240	1.390	1.270	0.534	2.9×10^4
7	240	1.392	1.481	0.479	3.0×10^4

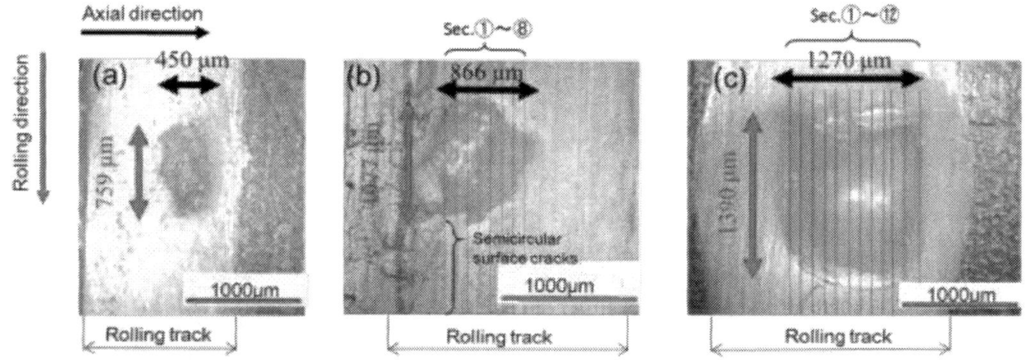

Fig. 3. The flaking on the rolling track of the PEEK shaft specimens: (a) 90 N, 4.5×10^6 fatigue cycles, (b) 120 N, 1.9×10^6 fatigue cycles, (c) 240 N, 2.9×10^4 fatigue cycles, and (d) 360 N, 2.8×10^3 fatigue cycles.

 https://doi.org/10.1051/matecconf/201926401004

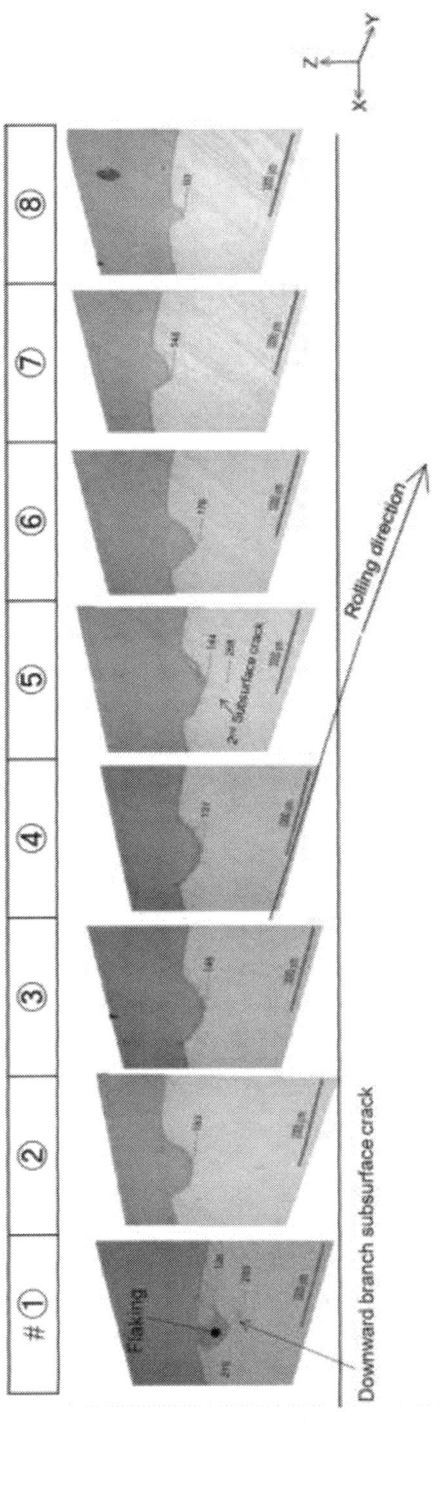

Fig. 4. 2.5D layer images of crosssection of the PEEK shaft rolling track at 120N, 1.9×10^6 fatigue cycles. Scale length is 1000 μm.

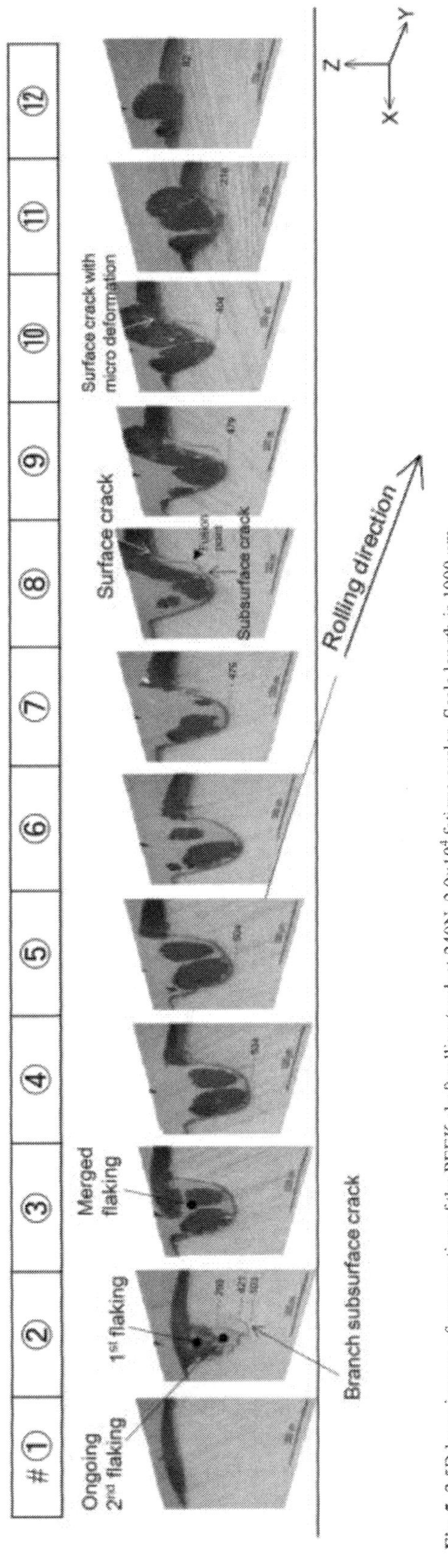

Fig. 5. 2.5D layer images of crosssection of the PEEK shaft rolling track at 240N, 2.9×10^4 fatigue cycles. Scale length is 1000 μm.

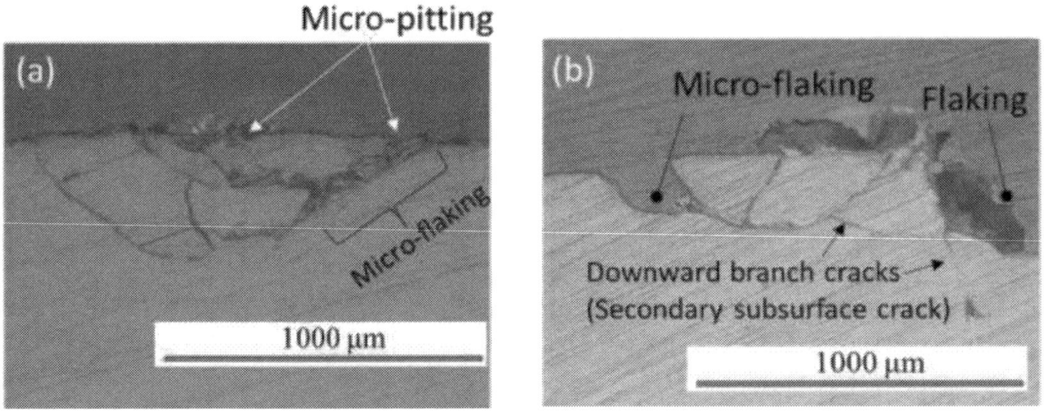

Fig. 6. Detail of the flaking origin at 120 N at 1.9×10^6 fatigue cycles; (a) micro-pitting, (b) micro-flaking and branch crack.

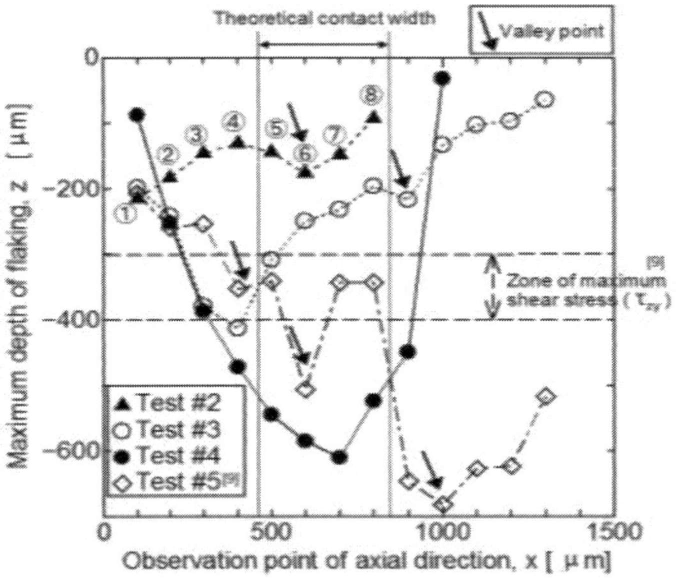

Fig. 7. Maximum depth of flaking of the PEEK shaft at 120 N.

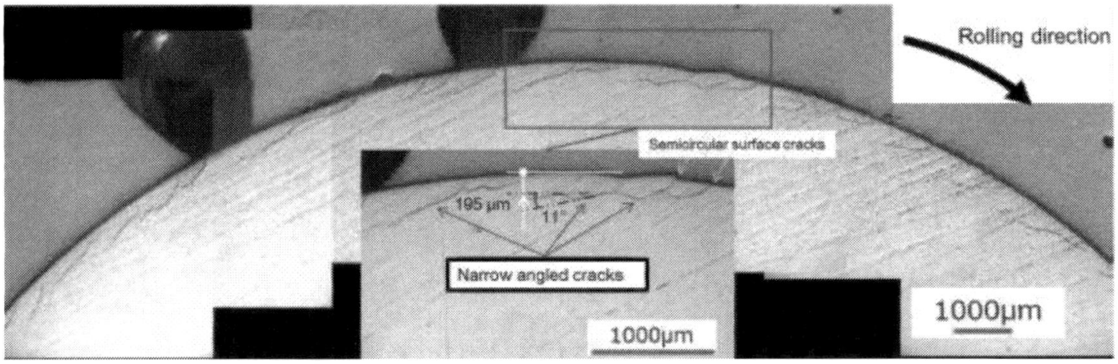

Fig. 8. Cross-section of near-surface cracks before fatigue fracture at 120 N (1.9×10^6 cycles).

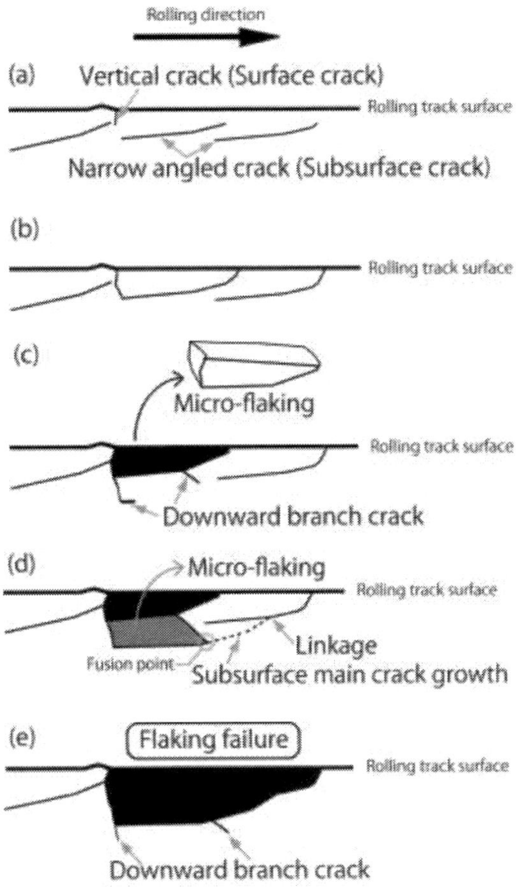

Fig. 9. Illustration of the flaking growth of PEEK shaft under one-point rolling contact fatigue.

3.3 Growth of flaking of PEEK shaft under one-point RCF

Figure 9 shows the flaking process near PEEK shaft surface. The flaking process is as follows:

(a) Narrow angled cracks occur near the rolling track surface due to internal shear stress.

(b) When the narrow angled cracks merge with each other, micro-flaking occurs.

(c) Downward branch crack propagates from the subsurface crack.

(d) Due to growth and fusion of subsurface cracks, secondally micro-flaking occurs.

(e) Subsurface main crack merges with other subsurface cracks (i.e. narrow angled cracks). Accumulation of the micro-flaking results the flaking failure.

Conclusion

In order to explore the flaking mechanism of PEEK shafts under rolling contact against alumina ball, one-point rolling contact fatigue (RCF) tests were carried out at 1200 rpm rotaion speed. The important results obtained from this work are summarized as follows:

(1) The length of the flaking was larger than the width. The subsurface crack propagated rapidly in rolling direction compared to in axial direction.

(2) Accumulation of micro-flaking results in the flaking growth. There were three types of subsurface cracks during flaking growth; narrow angled crack, downward branch crack, and main subsurface crack. The downward branch crack occurs from the subsurface crack, and propagates into the depth direction of PEEK shaft.

(3) Micro-flaking occurs due to the linkage of subsurface cracks and narrow angled cracks (and/or another subsurface crack) during the flaking process.

This research work is financially supported by JSPS KAKENHI Grant Number JP16K06042.

References

1. T. A. Stolarski, *Adv. Comp. Trib.edited by K. Friedrich*, Elservier, Chapter 17, (1993)
2. K. Friedrich, G. Theiler and P. Klein, *In: SinhaSK, BriscoeBJ, editors. Polymer Tribology*, (London UK: Imperial College Press, 2009)
3. M. Harrass, K. Friedrich, and A. A. Almajid, Trib. Int'l, **43**, 635-646 (2010)
4. K.Friedrich, Z. Lu and A. M. Hager, Theo Appl Fract Mech, **19**, 1-11 (1993)
5. K. M. Marshek, Wear, **52**, 141-146 (1979)
6. B.J. Briscoe, The tribology of Polymer Composite Materials, *Advanced in Composite Tribology edited by K. Friedrich*, Elservier, (1993)
7. A. Avanzini, G.Donzella, A.Mazzù, C.Petrogalli, Trib. Int'l, **157**, 22–30 (2013)
8. K. D. Dearn, S. N. Kukureka, and D. Walton, *In: SinhaSK, BriscoeBJ, editors. Polymer Tribology*, (London UK: Imperial College Press, 2009)
9. H. Koike, K. Kida, T. Honda, K. Mizobe, S. Oyama, J. Rozwadowska, Y. Kashima, and K. Kanemasu, Adv. Mat. Res., **566**, 109-114 (2012)
10. H. Koike, K. Kida, E. C. Santos, J. Rozwadowska, Y. Kashima, K. Kanemasu, Trib. Int'l, **49**, 30-38 (2012)
11. H. Koike, G. Yamaguchi, K. Mizobe, Y. Kashima, K. Kida, MATEC Web of Conferences, **130**, 9001(2017)
12. K. Kida, K. Ogura, The Society of Materials Science, Japan, **48, 10**, 1090-1094 (1999)
13. J. Rozwadowska, K. Kida, E. C. Santos, T. Honda, H. Koike, K. Kanemasu, Adv. Mat. Res.,**418-420**, 1613-1617 (2012)
14. H.Koike, K. Kida, K. Mizobe, X.Shi, S. Oyama, Y. Kashima, Trib. Int'l, **90**, 77–83 (2015)
15. B. J. Hamrock and D. Dowson, J. of Lubrication Tech **98(3)**, 375-381 (1976)

Influence of groove shape on rolling contact fatigue of PEEK-PTFE hybrid radial bearings in dry conditions

Masahiro Inagaki[1], Koshiro Mizobe[1], Katsuyuki Kida[1, a] and Yuji Kashima[2]

[1]*University of Toyama, Gofuku 3190, Toyama, 930-8555, Japan*
[2]*Kashima Plastic Bearings Corporation, 2-9-21 Himesato, Nishiyodogawa-ku, Osaka, 555-0025, Japan*

Abstract. In this study, rolling contact fatigue (RCF) tests were performed in order to investigate the effect of groove radius on the life of PEEK-PTFE hybrid radial bearings. Furthermore, solid state NMR spectroscopy was performed in order to investigate the molecular structure of PEEK. It was found that groove radius shape was related to limitations of the bearings, and the molecular structure of the PEEK was not changed by the temperature rise in this test. Furthermore, thermal failure of the PEEK was not affected by oxidation during RCF test.

1 Introduction

PEEK (Poly-Ether-Ether-Ketone) and PTFE (Poly-Tetra-Fluoro-Ethylene) have good mechanical features such as heat and corrosion resistance. In our previous studies on PEEK radial bearings [1-4], the effect of radial load on bearing life in dry conditions was investigated. Mizobe et al. found that the wear resistance of the PEEK hybrid thrust bearing with a PTFE retainer was superior to that without the PTFE retainer [5]. Koike et al. reported that the radial load capacity of the PPS-PTFE hybrid radial bearing was improved by the groove radius shape [6]. The failure pattern of these bearings was thermal deformation.

Solid state NMR spectroscopy is used for studies of polymers in order to investigate their properties [7-9]. For example, the solid state NMR spectra can examine the change in the molecular structure of polymer basing on the assignment of the carbon signals. However, the effect of rolling contact fatigue on the molecular structure of PEEK is unclear.

In this study, rolling contact fatigue tests were performed in order to investigate the effect of groove shape on bearing life of PEEK-PTFE hybrid radial bearings. Furthermore, solid state NMR spectroscopy was performed in order to investigate the effect of the rolling contact fatigue tests on the molecular structure of PEEK.

2 Test method

2.1 Specimens

Our bearing specimens (PEEK outer race, PEEK inner race, PTFE retainer, and nine alumina ceramic balls) were designed based on the Japanese Industrial Standards (JIS) B 1512-1 2011 (Rolling bearings - Boundary dimensions - Part1: Radial bearings) and 1521 2012 (Rolling bearings - Deep groove ball bearings, #6205). Figure 1 is a photograph of the bearing specimen. The outer diameter, inside diameter, and ball diameter were 52 mm, 25 mm, and 7.14 mm (9/32 inch), respectively. Two inner races were designed, with groove radii that were 101% and 112% of the alumina ball radius. The effect of contact radius on bearings features [10] and theoretical technique for calculation of fatigue life [11] were investigated for steel bearings. However, these techniques were not investigated for polymers. In this present work, two values of 101% and 112% are selected to investigate the effect of groove curvature on contact problem in polymer bearings. Figure 2 shows an illustration of the groove radius shape of the inner races. The inner race of 101% and 112% curvature radii have been labelled as "1.0r" inner race and "1.1r" inner race. Table 1 shows the material properties of PEEK and PTFE [12-14].

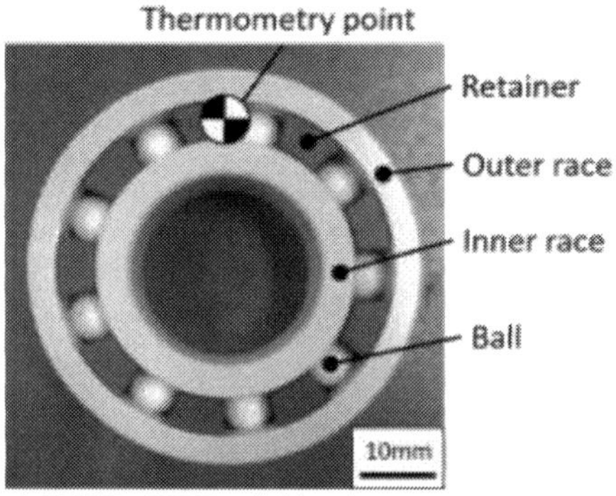

Figure 1. PEEK - PTFE radial bearing.

[a] Corresponding author: kida@eng.u-toyama.ac.jp

https://doi.org/10.1051/matecconf/201926401005

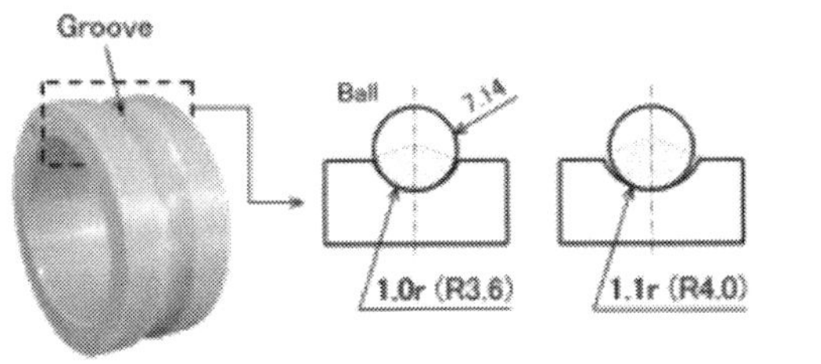

Figure 2. Curvature radius of groove of inner races.

Table 1. Material properties of PEEK and PTFE [12-14].

	Melting point [°C]	Glass transition point [°C]	Density [g/cm^3]	Tensile modulus [GPa]
PEEK	343	143	1.30	4.0
PTFE	327	115	2.16	0.4

2.2 Rolling contact fatigue (RCF) test

The RCF tests of PEEK-PTFE radial bearings were done using the RCF test machine which was developed by Kida's group [1-4,6]. Figure 3 shows a schematic illustration of the RCF testing machine. All tests were performed under dry conditions. Fatigue cycles were 1.0×10^6. The RCF tests were stopped when the specimens were broken by thermal deformation. Temperature and wear loss were measured using a thermal infrared sensor (Fluke Corporation, Fluke 59 Mini) and an electronic balance (A&D Company, BM-252).

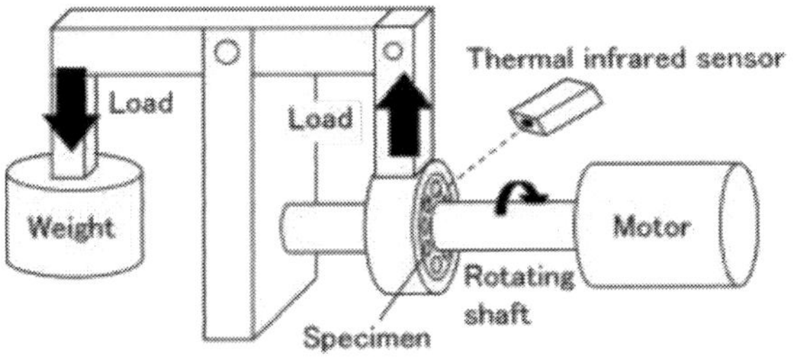

Figure 3. Schematic illustration of RCF testing machine.

2.3 Solid state ^{13}C nuclear magnetic resonance (NMR) spectroscopy

A solid state NMR spectroscopy (JEOL Ltd., JNM-ECX500) was used in order to investigate the effect of rolling contact fatigue tests on the molecular structure of the PEEK bearing. Table 2 shows the analysis conditions of the spectroscopy. A ZrO_2 file was used in order to assemble the powder samples.

Table 2. Analysis condition of solid state ^{13}C NMR spectroscopy.

Analysis method	CP/MAS
Rotational speed	10 kHz
Resonant frequency	125 MHz
Scan number	5000

3 Results and discussion

3.1 Relation between rotation speed and radial load

Figure 4 shows the failure conditions of the specimen. Failures are plotted as "×". They were caused by thermal deformation. Figure 5 shows photographs of the thermal deformation of the specimens. In the case of the thermal deformation, the inner race was melted more compared to the outer race. The thermal deformation was caused by the frictional heat. The load capacities of 1.0r and 1.1r specimens were 380 N and 476 N at 600 rpm. The 1.1r specimens did not break when the load was less than 284 N at 3000 rpm. The limit load of 1.1r specimens was higher than that of the 1.0r specimens. These results indicate that groove radius shape is related to the limitations of the bearings.

3.2 Wear loss

Figure 6 shows the relation between wear loss and radial load in four parts (inner race, outer race, ball, and retainer). In this test, wear loss of 1.0r was highest among all samples, whereas the wear loss of 1.1r specimens was less than that of 1.0r specimens. This indicates that the increase of the groove radius decreases wear loss.

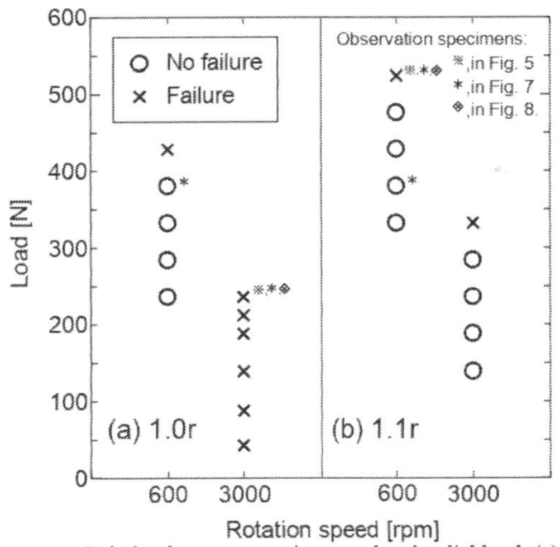

Figure 4. Relation between rotation speed and radial load. (a) is for 1.0r, and (b) is for 1.1r.

3.3 Thermometry

Figure 7 shows the relationship between temperature and the number of cycles. Temperature was measured until 1.0×10^6 cycles or thermal deformation. The thermal deformation occurred within 0.25×10^6 cycles when the temperature was over 60 °C. The feature operation temperature of the 1.1r specimens was similar to that of the 1.0r specimens. This indicates the thermal deformation occurred within 0.25×10^6 cycles irrespective of the groove radius.

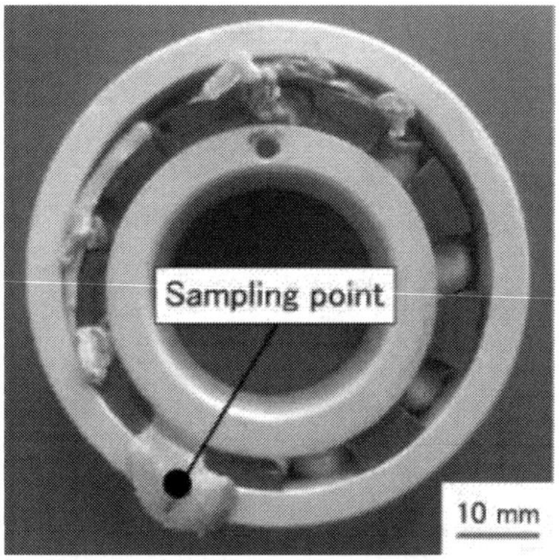

(a)At 600 rpm and 524 N

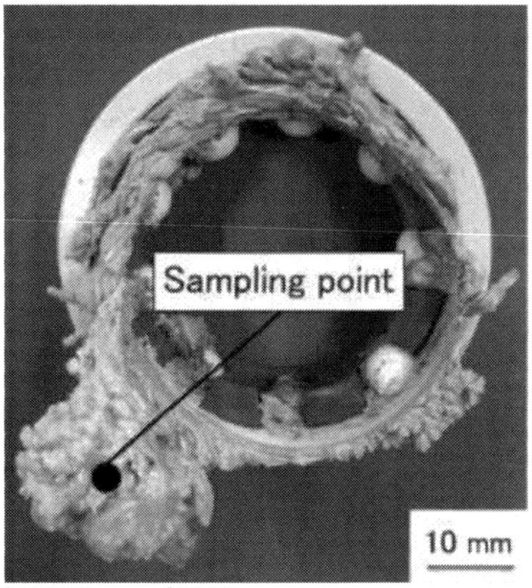

(b)At 3000 rpm and 236 N

Figure 5. Photographs of the thermal deformation specimens.

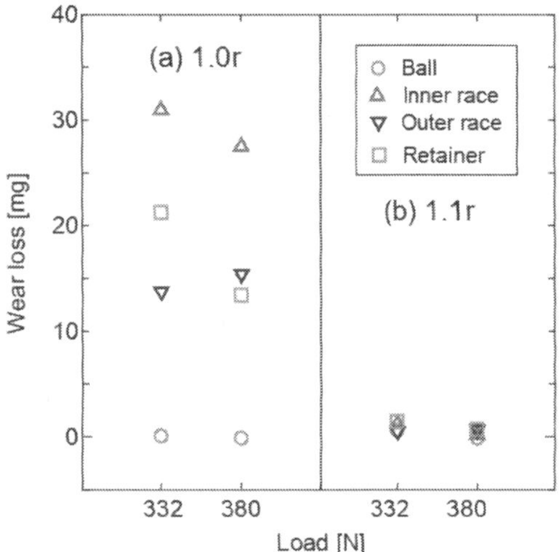

Figure 6. Relation between wear loss and radial load at 600 rpm. (a) is for 1.0r, and (b) is for 1.1r.

3.4 Analysis of solid state ^{13}C NMR spectroscopy

Figure 8 shows the solid state ^{13}C CP/MAS NMR spectra of PEEK bearings. Powder samples were assembled from the sampling points of Figures 5. Fig. 8 shows the spectrum before the test was similar to that after the test. All peaks can conform to the carbon positions in the literature [7, 8]. This indicates that the molecular structure of the PEEK is not changed by the temperature rise in this test. Therefore, it was found that the thermal failure of the PEEK bearing was not affected by oxidation during the RCF test.

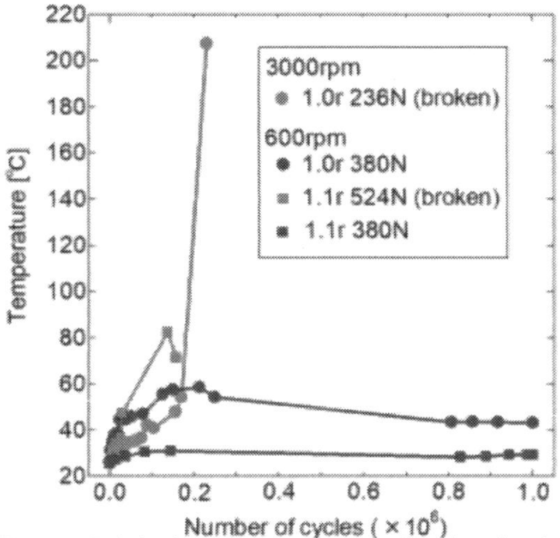

Figure 7. Relation between temperature and number of cycles.

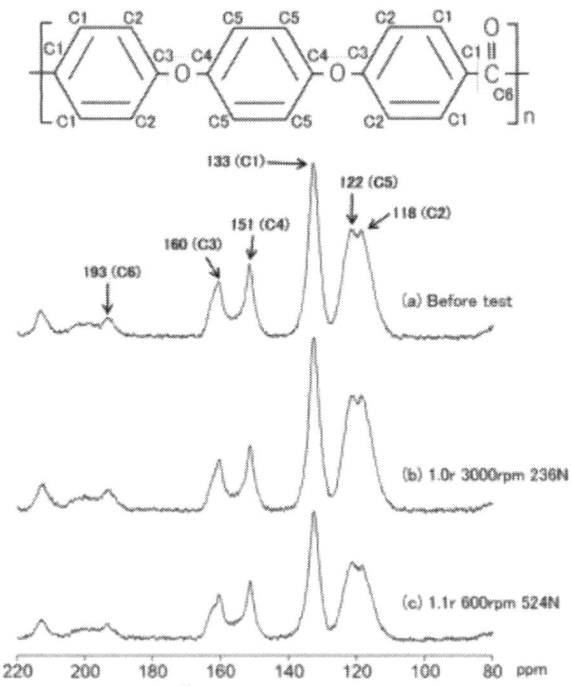

Figure 8. Solid state ^{13}C CP/MAS NMR spectra.

4 Conclusion

In this study, RCF tests and solid state NMR spectroscopy of PEEK-PTFE radial bearings were performed. The effect of radial load on life was examined. Furthermore, the effect of RCF tests on the molecular structure was studied. The following conclusions were obtained:

(1) Groove radius is related to the limitations of the bearings.

(2) Increase of groove radius decreases the wear loss.

(3) Thermal deformation occurs within 0.25×10^6 cycles irrespective of the groove radius.

(4) Molecular structure of the PEEK does not change due to temperature rise in this test.

References

1. H. Koike, K. Mizobe, S. Oyama, Y. Kashima, K. Kanemasu, K. Kida, Applied Mechanics and Materials **372**, 503-506 (2013)
2. H. Koike, K. Mizobe, S. Oyama, Y. Kashima, K. Kanemasu, K. Kida, Advanced Materials Research **683**, 385-390 (2013)
3. H. Koike, K. Kida, K. Mizobe, X. Shi, S. Oyama, Y. Kashima, Tribology International **90**, 77-83 (2015)
4. H. Koike, T. Matsumura, K. Mizobe, Y. Kashima, K. Kida, Materials Science Forum **878**, 142-147 (2017)
5. K. Mizobe, T. Honda, H. Koike, E. C. Santos, K. Kida, Y. Kashima, Advanced Materials Research **567**, 66-70 (2012)
6. H. Koike, K. Katsuyuki, T. Matsumura, K. Mizobe, Y. Kashima, MATEC Web of Conferences **130**, 09002 (2017)
7. Y. Paik, S. A. Chae, O. H. Han, S. Y. Hwang, H. Y. Ha, Polymer **50**, 2664-2673 (2009)
8. M. D. Poliks, J. Schaefer, Macromolecules **23**, 3426-3431 (1990)
9. J. Guenther, T. Bremner, J. Blümel, 1-9 (2012)
10. P. Romanowicz, B. Szybinski, Applied Mechanics and Materials **621**, 95-100 (2014)
11. P. Romanowicz, B. Szybinski, Applied Mechanics And Materials, **477-478**, 54-57 (2014)
12. Kashima Bearings, Inc., http://kashimabearings.com/
13. NSK Ltd., technical report **728h**, http://www.nsk.com/jp
14. Victrex plc., https://www.victrex.com/en/

A Comparative study of Wear and tool life of HSS TiN coated end mills and WC uncoated end mills.

Praphan Yawara [1,a] and Naphatara Intanon [1]

[1]*Faculty of Technical Education, Rajamangala University of Technology Isan Khon Kaen Campus, Khon Kaen, Thailand*

Abstract. The objective of this research is to compare the performance and economic value of TiN coated end mills and WC end mills on S45C slot milling. The commercial end mills that produce and provide by SAN Engineering and Supply Co., Ltd. and commercial S45C steels, were selected for this investigation. The models of wear were analyzed by the scanning electron microscope, SEM. The result showed that the wear models of the TiN coated end mills are abrasive and surface peeling, while the wear models of WC end mills are adhesive and uniform micro-chipping on the cutting edges. The analysis the economic value also represented that WC end mills are worth more than the others.

1 Introduction

A small and medium manufacture companies have an importance for manufacturing industries of Thailand. The cutting tools are an important part of the production process and represent a significant share in the costs of manufactured products. An evaluation of the operation process such as material tools and time is an important task which is one of the ways to reduce production costs.

In general, Tungsten carbide (WC) end mills were used for a machine carbon steel, and performs better at high speed level than High-Speed Steel (HSS) end mills, even though, the price of WC is relatively high. The Physical vapor deposition (PVD) is a coating technique that can improve a cutting performance and tool life of the tools [1]. The most common material that used to a coating to improve mechanical properties of the surface is Titanium. The workpiece that coated by titanium is used in the machine tools industry, automotive part and in the medical [2-4]. The TiN is wildly coated materials since it has the advantages properties include high hardness and adhesion, good ductility, excellent lubricity, high chemical stability and tough resistance to wear [5]. The wear of coated cutting tools has been analyzed by several researchers [6]. However, most of researchers are being studied in the extreme case such as high-speed milling, specific material or specific machine part. Due to the price of the coating process is very expensive and is wildly used in the laboratory. Thus, WC is mostly used in the area of SME industries even it costs more, however, it significantly shows an economic value better than a TiN coated end mills. Moreover, in recently the price of coated tools decreased to a possible level for the uses in industrial. Therefore, the SME should start to reconsider on the price and performance of coated tools for effective use again.

[a] Corresponding author: mppm345@hotmai;.com

This research is conducted in order to compare the performance and economy of TiN coated end mills and WC end mills to find a way of tools selection for Thailand SME.

2 Material and Methods

2.1 Work piece material and cutting tools

The materials of all experiments were the commercial S45C steels. (Chemical component in weight % : C 0.42-0.48, Mn 0.60-0.90, P 0.03, S 0.035, Si 0.15-0.35 Hardness 160-220 HB). There were two types of end mills used in this study. The first one is a TiN coated HSS end mills and the other one is WC end mills. Both of materials was supported by SAN Engineering and Supply Co.,Ltd

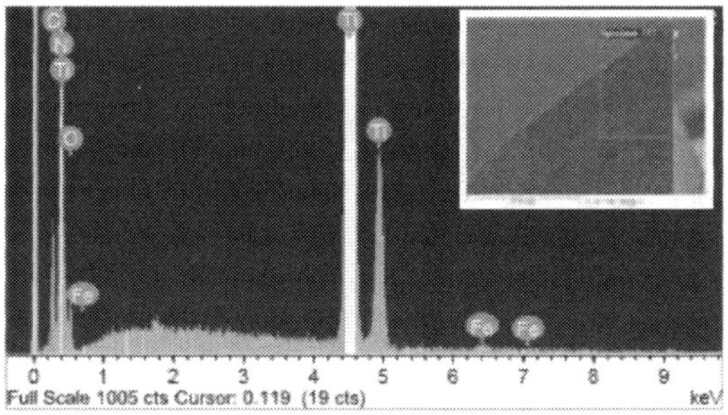

Figure1 Spectrum of TiN coated HSS end mill surface.

Table 1 shows the chemical composition of the TiN coated end mills surface. The atomic ratio of Ti and N is 25.89:43.86 or 1:1.69 while the typical TiN formation had roughly 1:1.

Table1. Chemical composition of TiN coated HSS end mills

Element	C	N	O	Ti	Fe
Weight %	7.30	26.84	11.31	54.17	0.38
Atomic %	13.91	43.86	16.18	25.89	0.15

The mechanical properties, especially the hardness, of the coated surface are upon the atomic ratio of Ti and N [7- 8]. The atomic ratio of Ti : N of TiN coated HSS end mills was 1:1.69, it had a higher than normal range value and the hardness of the coating surface is 1236.7 HV , lower than the general hardness of TiN about 30% . However, the variation of Ti: N ratio is a function of temperature [9] and therefore the hexagonal α-Ti phase could dissolve up to 15 atomic % nitrogen. [10].

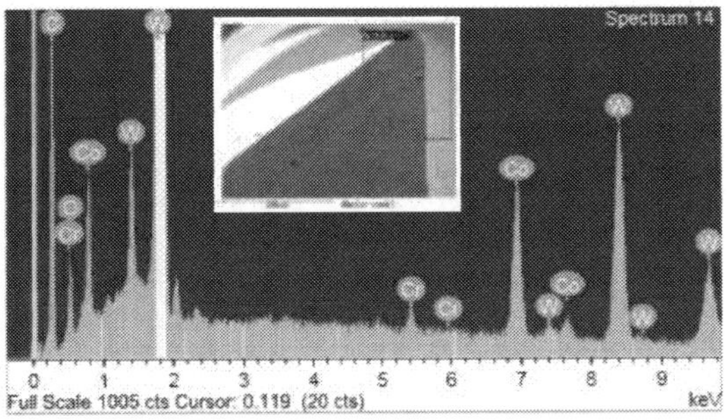

Figure2 Spectrum of WC end mills.

Table 2. Chemical composition of WC end mill.

Element	C	Cr	O	Co	W
Weight %	25.05	0.70	4.62	8.54	61.09
Atomic %	72.89	0.47	10.07	5.06	11.60

Table 2 showed that the atomic % ratio of W: C on WC end mills is 11.60: 72.89 or 1:6.28. The atomic ratio of typical WC was 1: 1, although the ratio of W: C was exchanged when added other materials depending to the manufacturer. In this case, the manufacturer added the cobalt (Co) into the chemical composition in order to increase the mechanical properties of the end mills.

2.2 Cutting parameters and procedure

Milling experiment was carried out under dry (non-lubricant) and wet (with lubricant) condition by using a Maike BSC 800 CNC millings machine. The optimization of cutting parameter, cutting speed (Vc) and feed per teeth (fz), for TiN coated HSS end mills on slot milling with a lubricant condition were 35 – 45 m./min. and 0.05 – 0.06 mm./teeth, respectively. And for the same condition recommended were 60 – 100 m./min and 0.02-0.04 mm/teeth for Vc and Fz, respectively [11].

$$\text{Spindle Speed} = \frac{1000 \times V_c}{\pi D} \quad (1)$$

For TiN coated HSS (Vc = 40 m./min)

$$\text{Spindle speed} = \frac{1000 \times 40}{\pi 12} = 1{,}061 \text{ rev. per min}$$

For WC (Vc = 80 m./min)

$$\text{Spindle speed} = \frac{1000 \times 80}{\pi 12} = 2{,}122 \text{ rev. per min}$$

However, this investigation has an economic comparison and parameters which used for the comparison are operation price and time (tool life). The operation time is upon the feed rate (vf). The equation of feed rate is $vf = f_z \cdot z \cdot n$ therefore, feed rate of TiN coated HSS end mills is $0.05 \cdot 4 \cdot 1061$ = 212.2 mm./min. and feed rate of WC end mills is $0.03 \cdot 4 \cdot 2122$ = 254.64 mm./min. Overall, the machine parameter of investigation can be show in table3.

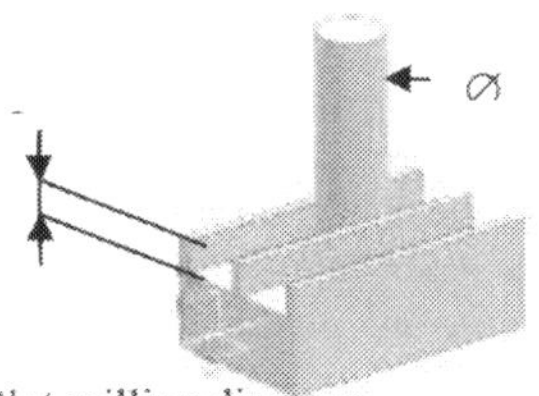

Figure3. Slot milling diagram

Table3. Machine Parameter

Work material	S45C	
Axial depth of cut	mm	6
Radial depth of cut	mm	12
Feed (f_z) (TiN coated HSS end mills)	mm/tooth	0.05
Feed (f_z) (WC end mills)	mm/tooth	0.03
Feed rate (TiN coated HSS end mills)	min./min.	212
Feed rate (WC end mills)	min./min.	255
Spindle speed (TiN coated HSS end mills)	rev./min	1061
Spindle speed (WC end mills)	rev./min	2122

The trial was two times repeated for each cutting condition in order to maintain the minimum error data. After finished in each condition, an end mills were cleaned and dried before diameter measurement. The measurement was taken two times follow a flute of end mills and average diameter value were recorded. The measurement tool of this research is a 0.001 mm. (0.0005 inch) digital micrometer (Mitutoyo)

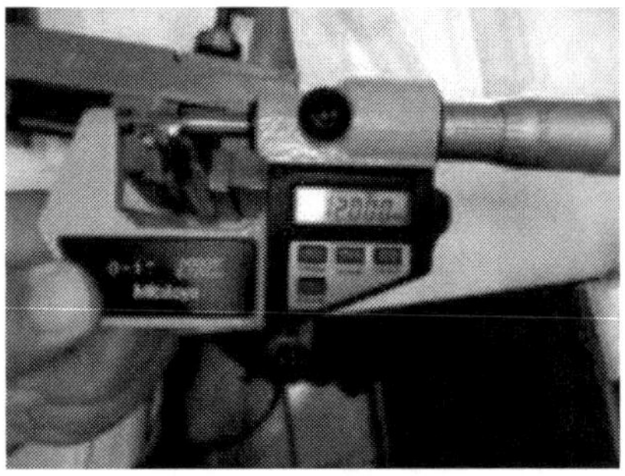

Figure4. Diameter measurement

3. Results

3.1 Wear and wear behavior

Table 4. and Figure 5. showed that the average of loss of TiN coated HSS end mills diameter after used in dry condition is .046 mm. and the average loss diameter of TiN coated HSS end mills in wet condition is .045 mm. and the average of loss of WC end mills in dry and wet condition are .0425 and .012 respectively. The diameter loss of TiN coated HSS end mills in both conditions are barely different, but it is rather different on WC end mills. Because of the wear mechanism of WC end mills in wet condition is an adhesive wear that lower wear rate than an abrasive wear in the dry condition. While the wear mechanism of TiN coated HSS end mills in the wet condition is micro-chipping that high wear rate as same as an abrasive wear that it is a wear mechanism of TiN coated HSS end mills in dry condition.

Table 4. Size of end mills

Type of end mills	Before (mm.)		After (mm.)	
	No.1	No.2	No.1	No.2
TiN coated (dry)	12.001	12.001	11.954	11.956
TiN coated(wet)	12.001	12.002	11.957	11.956
WC (dry)	12.001	12.002	11.958	11.960
WC(wet)	12.002	12.002	11.991	11.989

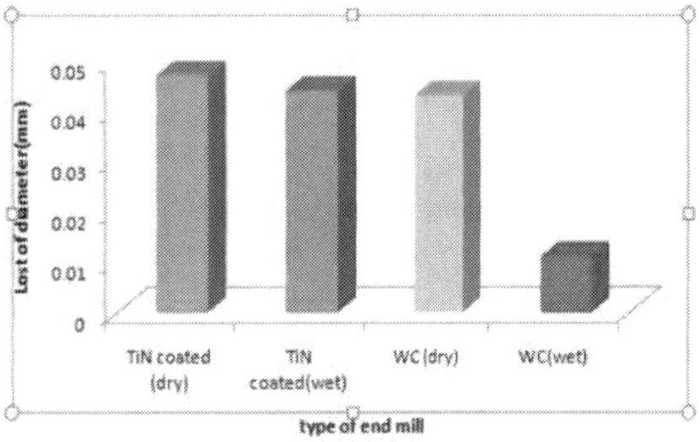

Figure5. Loss of diameter of end mills

The tool wear observed in the investigation is very specific. According to the optical and metallographic observation of wear behavior on the cutting edges could be established into two types of wear occurred in most cases: flank wear micro-chipping.

The initial observation of the wear behavior of both end mills on dry condition found that the rake surface was abrasive. The tool material (included coating material) loss from the tool flank which is normality of constant and extend over range of contact between tool and work piece. In wet condition, wear behavior was difference from dry condition. The rake surface of WC end mills was polished while the rake surface of TiN coated HSS end mills was not polished, but it showed non-uniform chipping (CH2). However, in the same condition (dry or wet) the machine parameter of tools was different, especially the spindle speed and feed rate.

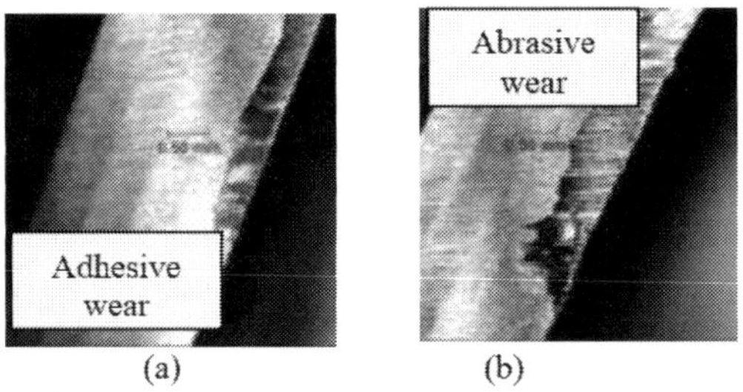

Figure6. The rake face of WC end mills (50X)
(a) wet condition (b) dry condition

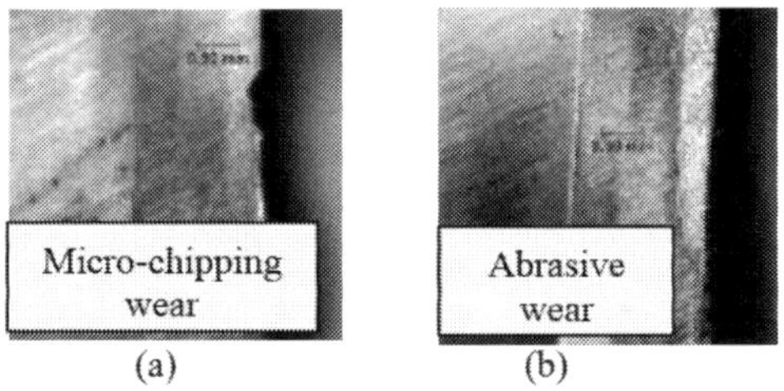

Figure7. The rake face of TiN coated HSS end mills (50X)
(a) wet condition (b) dry condition

In general, early stage of the tool wear was adhesive wear mechanical and next step is abrasive mechanical. However, it is difficult to separate due to most of wear were a combination of many mechanical.

To get a better insight into the wear behavior of cutting tools for all cutting edges, SEM was taken. The SEM images of TiN coated HSS end mills after milling is presented in figure 8.(a-d). The peeling of coating surface is prominent on dry condition (Figure.8(a-c)), conversely, on wet condition the peeling was decreased but the micro-chipping of the cutting edge was higher than a dry condition.

Figure 9. showed that the wear models for dry condition of WC end mills are uniform micro-chipping on the cutting edges (Figure 9. (a-c)). For the wet

https://doi.org/10.1051/mateconf/201926401006

condition, the wear models of is micro-chipping and abrasive wear (Figure 9. d).

3.2 Economic analysis

Tool life is a span of actual machining time by which a new tool can work until the wear of tool reaches a limit point. In general, limit point of end mills is flank wear, the recommended uniform flank wear is 0.3 mm. Form table 1. If the wear rate of end mills constants, the length of the slot of the TiN coated HSS can make in a dry condition as 3.26 m. and the length of the slot that the other end mills can make could be 3.33m, 3.53m and 12.5m. for TiN coated HSS end mills in wet condition, WC end mills in dry condition and WC in wet condition, respectively. The price of end mills that used in this investigation are THB 1450 ($44) for TiN coated HSS end mills and THB 1650($50) for WC end mills. The cost of end mills per meters of the slot is THB 444.67, THB 435, THB 467.6 and THB 132 respectively. The highest cost is WC end mills in dry condition; likewise, the lowest cost is WC end mills in wet condition.

4 Discussion

The wear behaviors of end mills are widely published. Sometime theoretically propounded findings were greater benefits to the manufacturers that could be the users of the tools. [8] In this study, the selected end mills were the commercials. The user's knowledge about the chemical composition of tools has limited. The user knows only that the main composition of tools which is the name of the type of tools and main composition of WC end mills is WC. In fact, WC end mills have many compositions and with some elements when is added into, it might change the hole properties. The WC end mills that used in this investigation has no common composition, the atomic % of W is low. But, it showed the higher performance than a TiN coated HSS end mills.

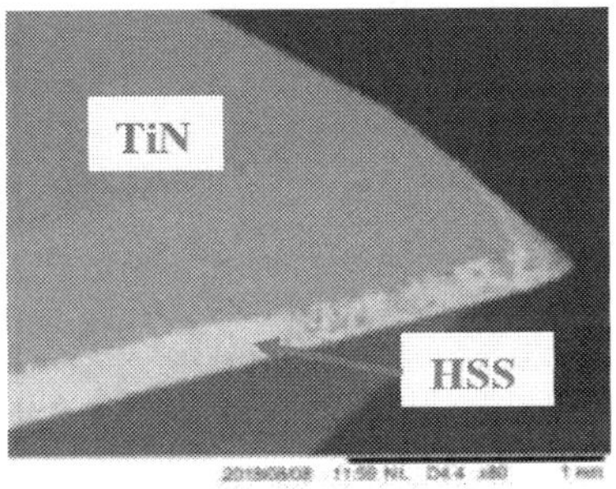

a.

b.

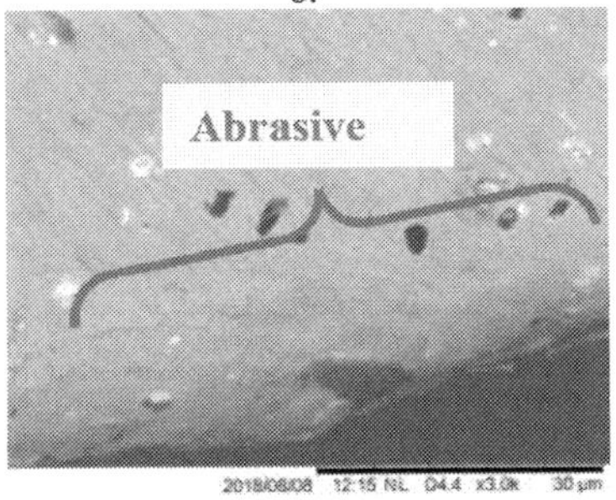

c

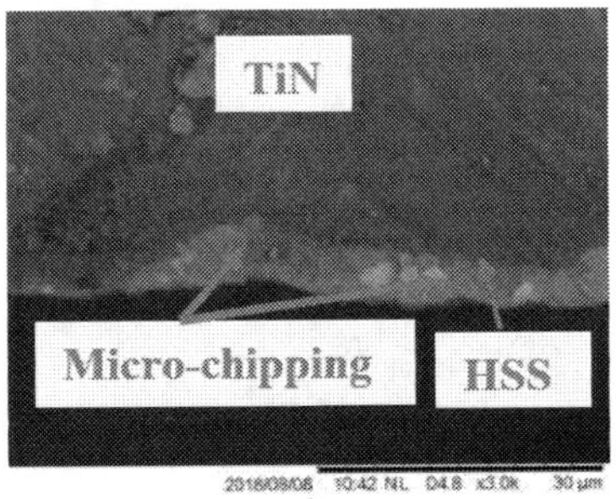

d.

Figure8. Cutting edge of TiN coated HSSend mills

(a)90X, (b)500X, (c)3000X (d) 3000X (wet)

However, it is not used to compare with the other WC end mills. So, the result could not be explained that it is better or worse than the other WC end mills that have a common composition. As same as, TiN coated surface has no common composition, but it showed a better wear resistant when used with lubricant.

In dry condition, the diameter of WC end mills and TiN coated HSS end mills after millings are almost equal. Because the TiN coated HSS end mills have a coated surface that high resistance and WC end mills was sensitive to the high temperature. Although, TiN has a high wear resistance, but after the coated surface peeling the wear, it will start and rapidly increase due to the wear rate of TiN coated HSS end mills is higher than the others. The contact area between tools and work piece in the slot mailing process are higher than end milling process and type of cutting is up cut are a good catalysts for peeling process. It is a main reason of high peeling of coated surface. In wet condition, the peeling of the coated surface was lower but the micro-chipping is greater than it, caused of the diameter after millings are almost equal. According to Sokovic,M. *et,al.* claimed that the coated end mills have no place in dry cutting. [11] As same as, WC end mills, the abrasive wear is lower but the micro-chipping wear is high. So, the loss of diameter is almost equal to the TiN coated HSS end mills and it is more different when compare to the end mills on wet condition.

The cost of tools is a main effect to the competitive advantage of SEM. This investigation shows that the best appropriate tool is WC end mill while the worst appropriate tool is also WC end mill, as well. (High price) It was upon the machine condition. Actually, the main jobs of SME was generally used steels, except the specific products that used a cast iron. So, the machine process was used with the lubricant and WC end mills should be selected. However, most of SEM orders are a job shop that the quantity of job is about 10 – 20 pieces. If the tools life is longer than the working time, the longer tools life (higher price) does not necessary. For example, from this investigation, working life of TiN coated end mills is 3.33 m. and 12.5 m. for WC end mills. If all of working distant of some job are 3.0 m., the SMEs should select the TiN coated end mills, but if they predict that they will make this job again in the next month, then WC end mills should be a good choice to be selected.

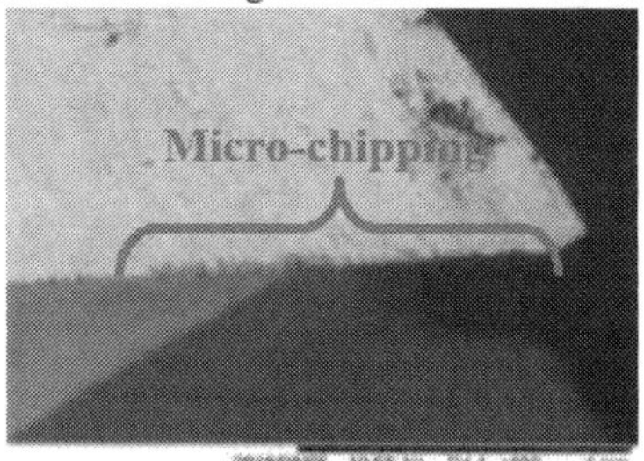

a.

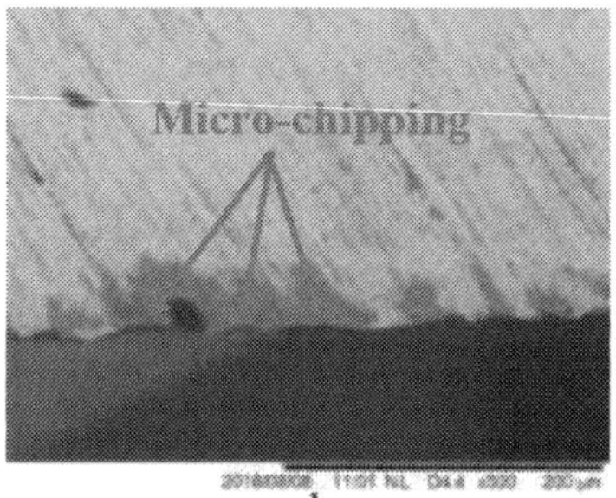

b.

c

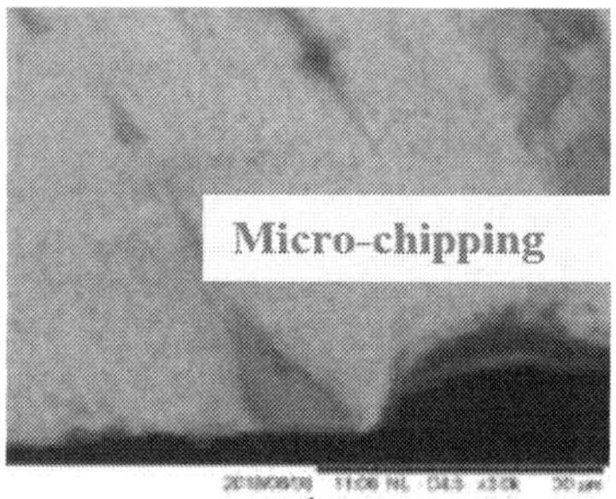

d

Figure 9. Cutting edge of WC end mills (a) 100X (b)500X (3)3000X (d)3000X (wet)

5 Conclusion

This investigation shows that the wear of end mills depending on the base material and machining parameters. According to the research of Sokovic, M. et, al. founded that the coated high-speed cutting tools showed quite different wear behavior depending on the base material used, type of coating and machining parameters [12]. The WC end mills has the advantage in wet condition even though it has high abrasion resistant and can also withstand higher temperatures. However, in the dry condition, TiN which was a ceramic material, it could perform better. Thus, the WC end mills is not appropriate in dry condition since without lubricant the temperature is high and might effect to wear mechanical of WC.

The economic value analysis found that WC is optimal for the wet condition but in dry condition should be used a TiN coated HSS end mills. Since the TiN coated surface has a good wear resistance and WC has higher hardness and it is a cause of micro-chipping in high-temperature.

6 Acknowledgements

This research was supported by The Faculty of Technical Education, Rajamangala University of Technology Isan campus, Khon Kaen. Thailand.

7 References

1. T. L. Banh, Q. T. Phan and D. B. Nguyen, ATM, **6[2]**, 244 (2004)
2. Y. Tamura, A. Yokoyama, F. Watari, M. Uo and T. Kawasaki, Mat.Trans., **43(12),** 3043 (2002)
3. D. Martinez-Martinez, C. Lopez-Cartes, A. Fernandez, J.C. Sanchez-Lope, Appl. Surf. Sci., **275**, 121 (2013)
4. X.H. Zheng, J.P. Tu, B. Gu, S.B. Hu, Wear **265,** 261 (2008)
5. W. Yanfeng, L. Zhenxian, W. Haonan, D. Jihong, Z. Changwei, Rare Metal Materials and Engineering, 46(5) 1219 (2017)
6. C.Y. Wang, Y.X. Xie, Z. Qin, H.S. Lin, Y.H. Yuan, Q.M. Wang, Wear **336-337 29** (2015)
7. J. Hoon Kang, K. Joo Kim, J. of Appl. Phys. **86**, 346 (1999).
8. U.W. Bläß, T. Barsukova, M.R. Schwarz, A. Köhlera, C. Schimpf, I.A. Petrusha, U. Mühle, D. Rafaja, E. Kroke, J. of the Euro. Cera. Society **35,** 2733 (2015)
9. K. Vasua, M. Ghanashyam Krishnaa K.A. Padmanabhan, Appl. Surf. Sci. **257,** 3069 (2011)
10. B. Subramanian, K. Ashok, M. Jayachandran, Appl. Surf. Sci. **255,** 2133 (2008)
11. Uddeholm corrax. Cutting data recommendations. www.uddeholm.com/files/Cutting_Data_Corrax_eng, (2007)
12. M. Sokovic, J. Kopac, L.A. Dobrzanski, M. Adamiak, J. Mater. Process Technol.**157–158**, 422. (2004)

NDE Detection and Characterization of Damage in Honeycomb Sandwich Composites

Kwang-Hee Im[1,a], Sun-Kyu Kim[2], Young-Tae Cho[3], Yong-Deuck Woo[1], Jong-An Jung[4] and David Kuei Hsu[5]

[1]*Department of Automotive Eng., Woosuk University 490, Hujung-ri, Samrae-up, Wanju-kun, Chonbuk, 565-701, Korea*
[2]*Division of Mech. System Eng., Chonbuk Natl. University, Duckjin-ku, Jeonbuk ,561-756, Korea*
[3]*Dept. of Manufacturing Design Eng., Jeonju University, Wansan-ku, Jeonju, Jeonbuk, 560-759, Korea*
[4]*Dept. of Mechanical and Automotive Engineering, Songwon University, 73 Sonarm-ro, Namgu, Gwangju 502-210, Korea*
[5]*Center for Non-destructive Evaluation, Iowa State University, Ames, Iowa 50011, USA*

Abstract. This study deals with a non-destructive evaluation technique for detecting defects in composite panels. In this non-destructive evaluation technique, various low-frequency techniques are applied such as a tapping technology. Another terahertz wave application will be utilized. In particular, the tapping technique is utilized in order to evaluate the characteristics of the honeycomb member based on the hysteresis effect analysis. Here, the area surrounded by the hysteresis loop in the force-displacement curve is related to the increased internal friction loss that is the cause of the energy absorption absorbed by the composite member after the load is applied. The loop area where the composite member was damaged was well agreed with the damage level. Artificial defects were manufactured on the surface of honeycomb sandwich composite panels. An NDE technique was proposed in order to detect defects by using terahertz waves and was discussed for tuning the practical use.

1 Introduction

Composite materials have the advantage of high specific strength and stiffness over compared to metals, but are brittle to impact damages. They could have their own unique damage and failure modes. In solid laminates, impacts can lead to delaminations at the ply interfaces[1]. In honeycomb structures, impact damage could make the structures fractured or crushed cores generated[2,3] and may also cause matrix cracking and delamination. The tap test imaging system can also be used to verify the internal substructures of composite parts. Tap test images are to be generated by both manual tapping and by using a mechanized scanner. The purshed system is used to produce images of both the impact duration of the taps and the deduced spring constant that shows the local stiffness of the structure [4,5,6]. These images determine quantitatively the degradation of stiffness due to flaws or damages and the enhancement of stiffness due to internal reinforcements, core splices and potted core. Structures used in the experiments included actual honeycomb sandwich panels and GFRP solid composites[7,8].

As an antoher tecnique, technology and instrumentation in terahertz radiation have provided a probing field on the electromagnetic spectrum because the terahertz radiation has a shorter wavelength, relatively higher resolution than microwaves, and lower attenuation [9,10]. Also the terahertz time domain spectroscopy (TDz-TDS) is leading noncontact accurate detection of flaws and impact damages in composites, in which the TDz-TDS is based on photoconductive switches, which rely on the production of few-cycle terahertz pulses using a femtosecond laser to excite a photoconductive antenna [2]. In this study, terahertz waves are explored as an NDE tool for detecting and characterizing flaws and damage in GFRP solid composites [11].

When GFRP solid laminates and honeycomb sandwich composite structures are subjected to impact damages, their stiffness would be influenced to a degree of severity in impact. In particular, the contact stiffness in the damage zone could decrease considerably. In addition, the resultant internal damage can produce a hysteresis loop in the load-displacement response of the impact site. This approach shows the stiffness reduction and the mechanical hysteresis caused by impact damage in composites and strives to develop nondestructive inspection methods based on these changes. Since qualitative manual tap test is perhaps the most widely practiced inspection for composites in service, particular attention is given to the development of tap tests as a quantitative nondestructive evaluation method based on the extraction of stiffness and mechanical hysteresis data from the force-time history of a tap test. Addtionally the use of T-rays as an NDE tool for detecting and characterizing flaws and impact damage in GFRP solid composites will be investigated compared with the results of tapping testing.

[a] Corresponding author: khim@woosuk.ac.kr

2 Experimentation

2.1 Contact stiffness and tap test

In the inspection of composites for impact damage in GFRP solid laminates or disbond in sandwich panels, manual tap tests were widely used. An impact-damaged or disbonded area would make a dull or hollow sound produced ; whereas an undamaged region would make a crisp and solid sound produced. Such hearing-based tap tests are dependent on operators and will be easy is subjected to the interference by ambient noise. Instrumented tap test devices are still available for many years. We purchased the Computer Aided Tap Tester (CATT) from Iowa State University. These instrumented tap testers all made use of the contact time between the tapping mass and the surface to be inspected. The damage caused by an impact can reduce the stiffness of the part and lengthen the contact time considerably [3].

Figure 1. The components of the tapper system schematically.

The advantage of the mechanical tapper is that the tap spacing in the scan direction is highly constant compared to that of a hand scan, and that it is much less tedious.The use of the mesh template had very negligible effect on the duration during impacting.

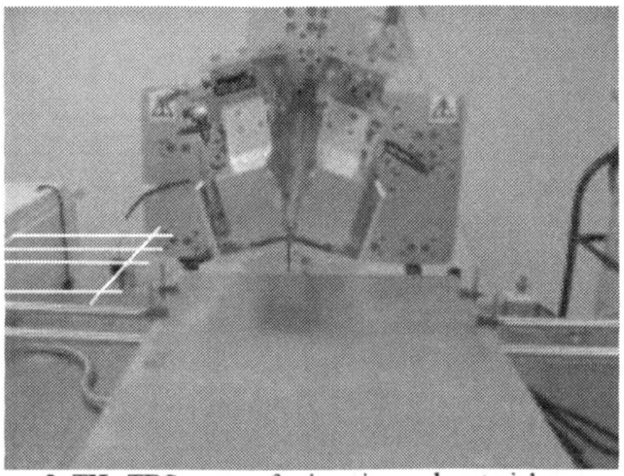

Figure 2. THz TDS system for imaging and material parameter measurement

2.2 T-ray scanner

Figure 2 shows a schematic of T-ray time-domain spectroscopy system (THz-TDS), the nondestructive test system. As shown in Fig. 2, a reflection mode was ste up for scanning defects on the honey-comb sandwich panels. The T-ray instrumentation systems used in this research were provided by TeraView Limited in England. The instrumentation is consisting of a time domain spectroscopy (TDS) pulsed system and a frequency domain continuous wave (CW) system.

The TDS system has a frequency range of 50GHz – 4 THz and a fast delay line up to 300ps. The beam is focused to focal lengths of 50 mm and 150 mm and the full width at half maximum (FWHM) beam widths are 0.8mm and 2.5mm respectively. The TDS system can be configured for through-transmission or reflection (small angle pitch-catch) measurements. The frequency range of the CW system is 50GHz – 1.5THz, with the best resolution being 100MHz. The focal lengths of the CW system are also 50mm and 150mm. Both the TDS and the CW systems are fully fiber optics connected[11].

2.3 Relationship between stiffness and contact time

The several recent extensions of the Hertzian contact theory have confirmed the impact time (i.e., contact time) between two impacting objects. For example, Greszczuk [4] analyzed the impact time of a spherical impactor dropped onto a flat plate and contacted on an expression for the impact time as a function of mass and radius of the impactor, the drop height, and the elastic properties of the target and the impactor. Such meanings were that the contact time of a tap is less sensitive to the velocity or the force with which the tap was made. Figure 3 shows the force-time history of taps made with the repetition tapping on the composite honeycomb panel with the differnet height. An accelerometer was utilized and its weight is 14.16g with different heights of 5mm to 8mm.

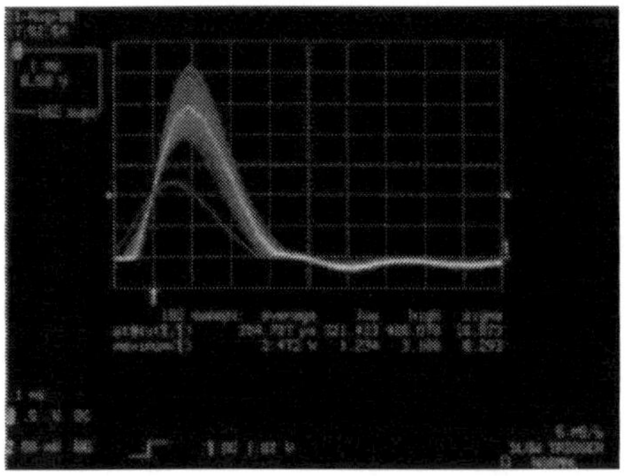

Figure 3. Force-time history of taps on a composite honeycomb panels with the same repetition tapping. [Horizontal axis 100μs/div, vertical axis 0.5V/div].

It was found that the contact time is approximately the same even if the different tapping force loaded on the panel. The curves were the voltage output of an accelerometer, which meant the tapping mass. It was shown that the GFRP composites and honeycomb sandwiche panels may be modeled on the basis of only the stiffness (spring constant) of the structure and the mass of the tapper [6].

Notice that the force-time history may be considered for one half of a sine wave cycle. So, the contact time τ is $1/2f$ and π/ω due to $f=1/\tau$ and $\tau=2\ \pi/\omega$. A spring constant k is follows [5,6]:

$$k = m\,(\pi/\tau)^2 \qquad (1)$$

In this manner, the spring constant k can be solved from the obtained tapper mass m and the measured contact time τ. Several tapping measurements have shown good agreement between the values of k deduced from tap test and measured in actual static load tests for damaged and undamaged honeycomb sandwich panels and GFRP composite laminates.

3 Results and discussion

3.1 Impact damages

By using TDS T-ray waves, the several delaminations in GFRP solid composite laminates were imaged based on the reflection mode from on the power described above. The GFRP sample was a 24-ply woven glass epoxy solid laminate and the sample was impacted with a 51mm diameter tup and an impact energy is 16.6 Joules. The C-scan images were obtained from the impact point using reflection mode. Figure 4 (a) shows the photo with the size of impact point. Figure 4 (b) shows the T-ray scan image in order to produce the images of the impact-induced delaminations with the time gate range of 43-69ps. Figure 4(c) and (d) show an impacted images with the different time gate range of 47-69ps and 56-69ps respectively. These images clearly are showing the decreasing size of the delaminations toward the impact point.

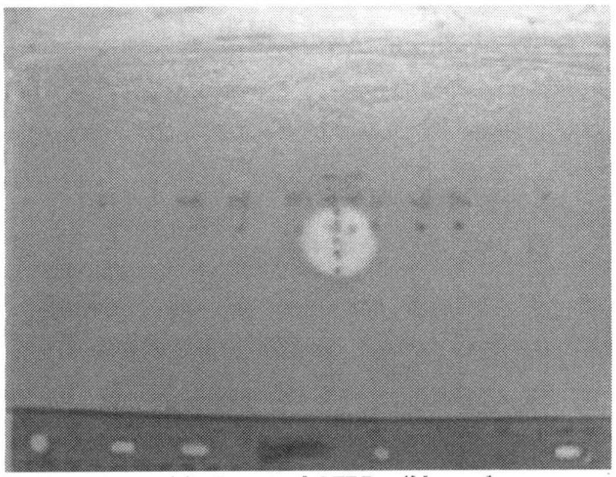

(a) Impacted GFRP solid sample

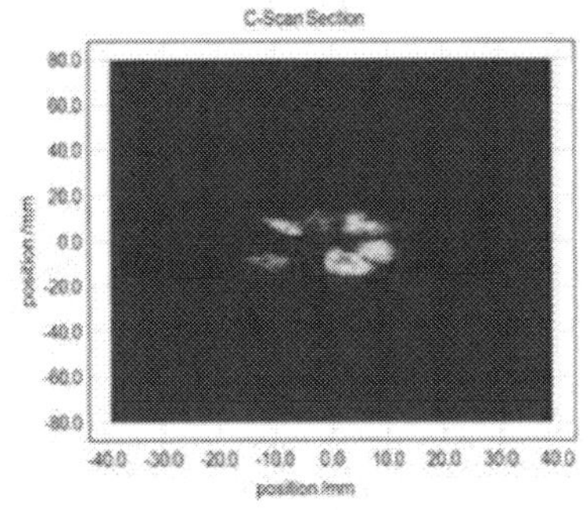

(b) 43-69ps

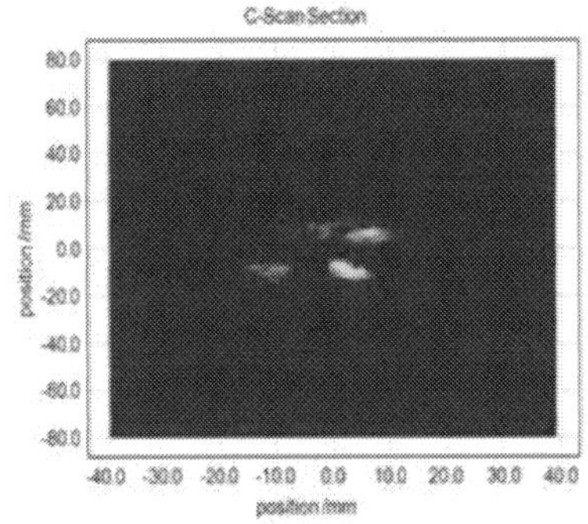

(c) 47-60ps

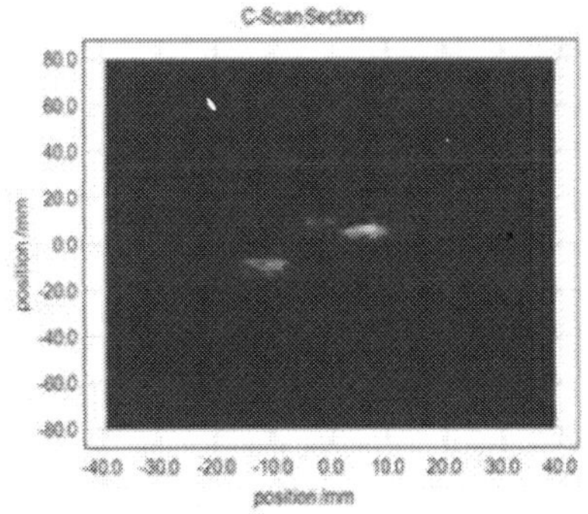

(d) 56-69ps

Figure 4. Time domain TDS waveform of GFRP composite laminate and impact delamination images using different time gate (43-69 ps, 47-69 ps and 56-69 ps in order)

The impact damage threshold had been investigated of GFRP solid laminates using the Instron impact tester Dynatup 8200 (such data are of interest to the work on solid laminates in Sandia) and the NDE of the damages so produced. In addition, we have also investigated the acoustic spectrum of the impact in a tap test. This was motivated by several factors. First, the tap test of a light-weight, flexible structure (e.g., honeycomb sandwich) will encounter limitations on a thick solid laminate, as the nature of the test transitions from a local test to a global test. It is therefore relevant to explore the acoustic spectrum.

We have founded that the effect of the tapping mass were investigated for detecting defects in GFRP solid laminates and found that the common wisdom of a "bigger hammer" will probe deeper did not seem to hold

https://doi.org/10.1051/matecconf/201926401007

true. Figure 5 shows the tap test locations on a 12.7mm thick GFRP laminate (tapping through 14.3mm of material). With the consistent taps produced by the recently developed motorized tapper, we can now make use of the contact force as well as the contact time in a tap test.

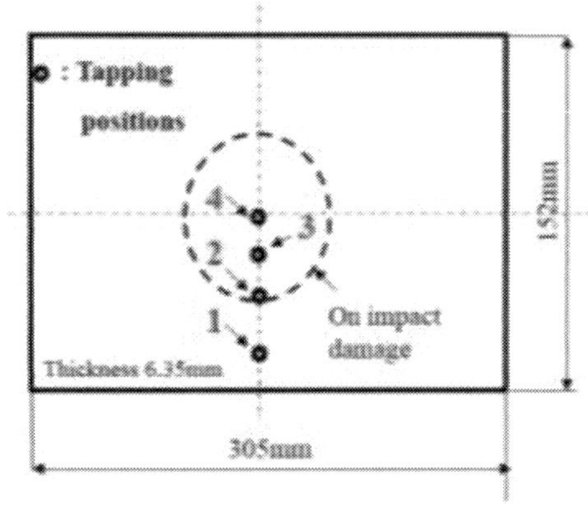

Figure 5. Contact time effect of tapper mass on impact damage

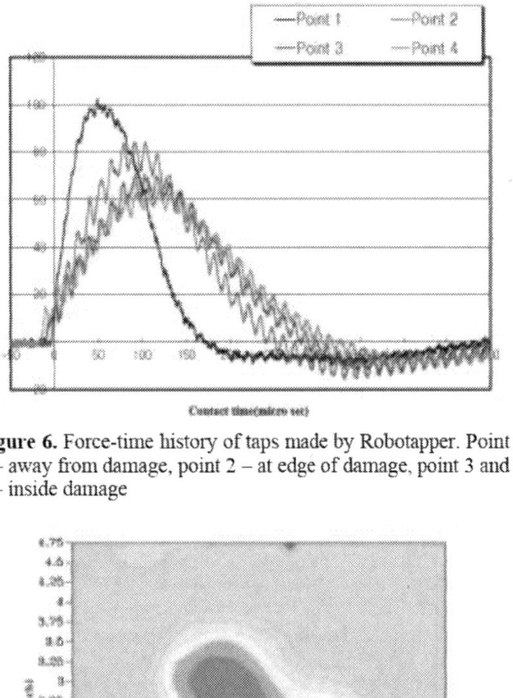

Figure 6. Force-time history of taps made by Robotapper. Point 1 – away from damage, point 2 – at edge of damage, point 3 and 4 – inside damage

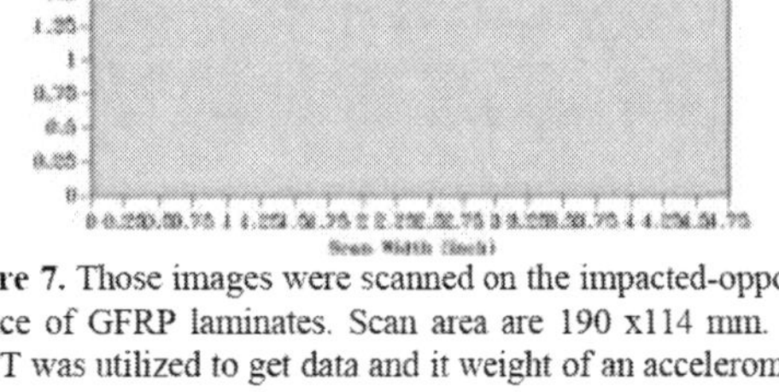

Figure 7. Those images were scanned on the impacted-opposite surface of GFRP laminates. Scan area are 190 x114 mm. The CATT was utilized to get data and it weight of an accelerometer is 14.16g.

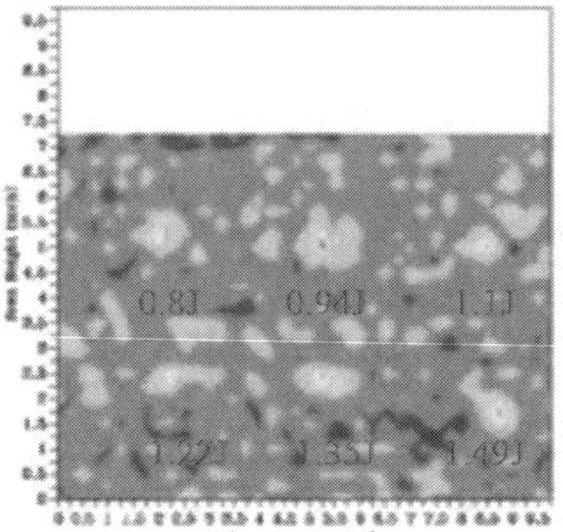

(a) Contact time

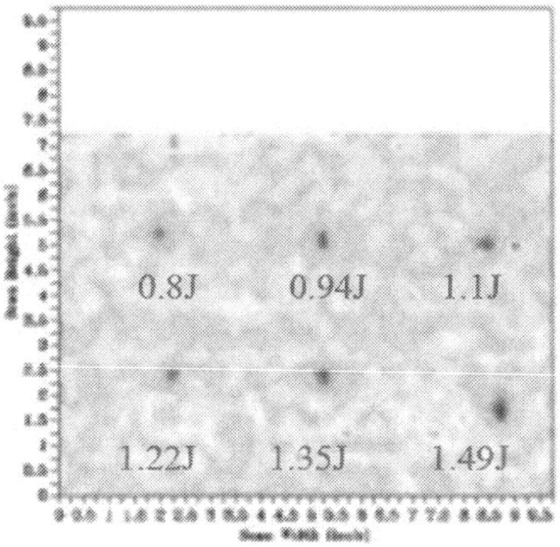

(b) Stiffness

Figure 8. Scan image of impact damages on a glass composite honeycomb panel based on the tapping test.

Figure 6 shows the force-time history of taps made by the tapper. Point 1 is at a location away from an impact damage on a ~6.35mm GFRP laminate, point 2 is on the edge of the impact damage, and points 3 and 4 are within the impact damage. As expected, the peak force was lower in a less-stiff, damaged area and that the contact time "tau" is longer in the damaged area. With the tapper, it is now possible to acquire the entire tap response and use both the contact force and the contact time in the detection and quantification of flaws and damage. The specimen used for Figure 7 containing the pre-existing impact damage show the tapping image. Here,GFRP skin honeycomb sandwich panels were tapped by an instrumented tap test system (CATT: Computer Aided Tap Tester), which was developed by Iowa State University for evaluating the impact damages of GFRP solid laminates. The tapper scans were made for both sides of impact-damaged sample. Figure 7 shows the different impact energies of 0.8J, 0.94J, 1.1J, 1.22J, 1.35J and 1.49J. Figure 7 (a) and (b) show the tapped image based on the contact time and stiffness respectively. The two scan images seem to be similar to each other. However, the degree of image varied with the degree of severe impact damage of the honeycomb sandwich panels.

Here τ and k-based images of the samples could easily be obtained by this CATT software. It is found that stiffness of impacted areas is pretty lower than other area as expected. Also, a similar trend was observed for the contact time as Figure 7 as expected.

Therefore for enhancing good imaging and quantitative capabilities, this tap test is an attractive NDE testing tool for inspecting GFRP solid laminats and especaily sandwiche penels. It was recommended the amount of NDT information obtained per unit cost is believed to be among the highest of all the NDT techniques because of the simplicity. Also, the T-rays could be useful for various NDT applications. However, their applicability is currently limited by the high cost of instrumentation and difficulty of use. We are expecting the growing demand will cost down for the easy-to-use.

4 Conclusions

This paper showed several advances in doing the tap test a more quantitative technique for imaging, detecting and characterizing the defects and impact damage in composite structures. Even if there are various NDE methods available, the tap test is the simplest, less costly, and most widely experienced mode of inspection for composites, especially GFRP solid laminates and honeycomb sandwiches. For this reason, the capabilities of tap test could lead to a broader impact in the NDE of composite materials. Also the non-conducting composites with glass and fibers, the T-ray radiation can complement ultrasonic NDE, especially with its penetration degree and regardless of shadow effect. The applicability of T-ray spectroscopy for materials evaluation, contamination detection, and fluid ingression was studied. By the way, it was found that the T-ray waves were very sensitive to water and oil as well as impact damages.

5 Acknowledgment

This research was supported by Basic Science Research Program through the National Research Foundation of Korea (NRF) funded by the Ministry of Education, Science and Technology (No.2018R1D1A1B 07049775 and 2015R1D1A3A01016775).

6 References

1. C. F. Buynak and T. J. Moran, Review of Progress in Quantitative NDE, **6B**,1 (1987)
2. D. K. Hsu, *American Society for Composites, 21st Technical Conference*, **1**, 1 (2006)
3. D. K. Hsu, D. J. Barnard, and D. P. Roach, *Materials Evaluation* (American Society for Nondestructive Testing 2009)
4. L. B. Greszczuk, *Impact Dynamics*, (John Wiley & Sons, Inc., 1982)
5. D. K. Hsu, D. J. Barnard, J. J. Peters and V. Dayal, Rev. Prog. Quantitative NDE, **19**,1 (2000)
6. Cory D. Foreman, *M.S. Thesis* (Iowa State University 2008)
7. Cory Foreman, V. Dayal, D. J. Barnard, and D. K. Hsu, Rev. Prog. Quantitative NDE, **28**,1 (2009)
8. P. Cawley and R.D. Adams, J. Sound and Vibration, **122**, 2 (1988)
9. K.H., Im, K.S. Lee, I. Y. Yang, Y. J. Yang, Y. H. Seo and D. K., Hsu, Int. J. of Prec. Eng. and Manuf. **14**, 6, (2013)
10. R. Huber, A. Brodschelm, A. Tauser, Appl. Phys. Lett., **76**,1 (2000)
11. J. V. Rudd, and D. M. Mittleman, J. Opt. Soc. Amer. B, **19**, 2, (2000)

Heat Treatment Effect on Mechanical Properties of 3D Printed Polymers

Eduardo de Avila[1], Jaeseok Eo[2], Jihye Kim[1] and Namsoo P. Kim[1,2,a]

[1]*Center for Printable Materials Certification (CPMC), The University of Texas at El Paso, El Paso, TX 79968, USA.*
[2]*Metallurgical, Materials and Biomedical Engineering Department, The University of Texas at El Paso, El Paso, TX 79968, USA.*

Abstract. *PMMA, PC, and PEEK are thermoplastic polymers that possess favorable properties for biomedical applications. These polymers have been used in fields of maxillo-facial, orthopedic, intraocular surgery, and bio-implant, due to their excellent mechanical properties, osteoinductive potential, and antimicrobial capabilities. In this study, the effect of heat treatment on the mechanical properties of 3D printed polymers was characterized. By modifying printing temperature and post heat treatment process, the mechanical properties were specifically tailored for different applications, correlating with the properties of the implants that are commonly made using molding processes.*

1 INTRODUCTION

Mechanical properties of materials including elasticity, yield strength, ultimate tensile strength, flexural strength, and ductility are part of materials specifications, and these properties determine the duration, biocompatibility, regeneration of biomaterials when the implantation is applied. Materials scientists have studied various ways of modifying the intrinsic properties of materials, either by combining one or more materials to produce different types of alloys or composites or by subjecting them to specific processes, such as heat treating. Heat treatment of polymers is one of the most effective methods of modifying the mechanical properties and widening their applications.

Additive Manufacturing (AM) or 3D printing technology have been applied to a biomedical application for the *in-vitro* and *in-vivo* test using metal/polymer implantation materials. However, the materials and process itself has not been standardized so far and still has unknown factors [1]. 3D printed polymers or composite polymers for medical application need to be tested for printability, reliability, and reproducibility in the geometry required for tensile specimens according to ASTM D638 [2]. The uncertainties of 3D printing technology which come from the heat transformation, energy dissipation and cooling should be clarified. For the reliability test, the specimens were heat treated at various temperatures to measure the changes in mechanical properties. The heat treatment of polymer-based composites resulted in improvement of mechanical properties, including fatigue resistance, fracture toughness, tensile strength, wear resistance, and friction coefficient [3]. The flexural properties have been tested following ASTM 790 [4]. The friction tests in polymers have reported to be affected by various aspects such as real contact area, which is dependent on temperature and sliding velocity [5].

Velocity has a different effect on friction characteristics, depending on whether the test is conducted at low or high speeds. This is also applicable for the contact pressure or applied load [6]. In this study, we have attempted to compare the friction behavior of the polymers before and after heat treatment.

Thermoplastic polymers such as PMMA, Polycarbonate (PC), and Poly(aryl-ether-ether-ketone) (PEEK) have been used for the mechanical analysis.

PMMA is a non-degradable polyacrylate, the most commonly applied non-metallic implant material in orthopedics. Due to its excellent biocompatibility, hemocompatibility, and ease of manipulation, PMMA has been widely used in many medical devices, including blood pumps and dialyzers. It is an ideal candidate material for implantable ocular lenses and hard contact lenses by its optical properties [7].

Polycarbonate (PC) is one of the most popular engineering resins in the medical device market, offering an unusual combination of strength, rigidity, and toughness. It provides glass-like clarity that is critical for many medical devices [8]. It is a thoroughly tested plastic used today and has a safety track record of more

[a] Corresponding author: nkim@utep.edu

than 50 years. It is available in different grades that comply with biocompatibility testing standards[8].
Poly(aryl-ether-ether-ketone) (PEEK) belongs to a family of high-temperature thermoplastic polymers, consisting of an aromatic backbone molecular chain, interconnected by ketone and ether functional groups. PEEK has been applied as a biomaterial over a decade, and a half of the uses are successful in the clinical history of load sharing fusion applications in the spine [9][10].

Biomaterials, human-made or natural origin, are used to supplement, direct, or replace parts or functions of living tissues in the human body. Biomaterials perform in the body's internal environment, withstanding the pH of body fluids as well as stress and compression forces, which may vary up to 10 times the body weight, as when jumping (9,000 N) [11].

In general, metals or ceramics are used for hard tissue applications, and polymers are used for soft tissue applications [12]. A wide number of polymers are used in various biomedical applications; however, most of the plastic composites are too flexible or too weak to meet the demands of a particular applications. One of the attempts to improve the mechanical strength and electrical conductivity of 3D printed structures was implemented by adding nanoparticles to polymers, metals, and ceramics matrix. [13-15].

It is reported the tensile strength of the biopolymer composite could be increased by 40%, due to the increase of the interface area, packing factors and internal energy. [16] The feasibility of the printability and modification of specific properties by heat treatment is explored to evaluate the practicability of polymer, polymer-composites and dispensing powders in non-traditional applications [17-20]. In a case of 3D printing with polymer or composite materials, effects of the apparent density changes in various printing techniques, including theoretical studies on electrical conductivity as well as mechanical strength were discussed, and standardization of the mechanical strength studies have also been reported. [21]. The Hagen–Poiseuille (HP) equation was applied to find the constant volume flow rate and the variation in the amount of discharged fluid for non-Newtonian material and applied to 3D printed liquid drugs. A procedure to control the constant volume flow has been formulated into 3D-printed structures and composite for the pharmaceutical training program at The University of Texas at El Paso (UTEP) and has been patented [22][23].

2 MATERIALS AND HEAT TREATMENT

All samples were printed as filaments having a 1.75×10^{-3} m diameter using 3D printers designed by the research team, Center for Printable Materials Certification (CPMC), with a nozzle diameter of 8.0×10^{-4} m. The filaments used in this study were: PC-Plus from Polymaker with a recommended printing temperature (RPT) of 250 – 270 °C; PMMA from Rigid Ink, with a RPT of 250 – 260 °C; and PEEK Filament from Victrex with a RPT of 360 °C. Tensile and bending test samples were printed according to ASTM DS 638 standard type V, and ASTM 790 (see Figure 1), and tested in Test Resources 100 series equipment. Friction coefficient was measured on discs with a diameter of 4.5×10^{-2} m, using an Anton Paar TBR tribometer, with a 6.0×10^{-3} m ball, a 5 N load, 30 rpm, and for 20 minutes. All tests were performed at room temperature ≈ 20 °C.

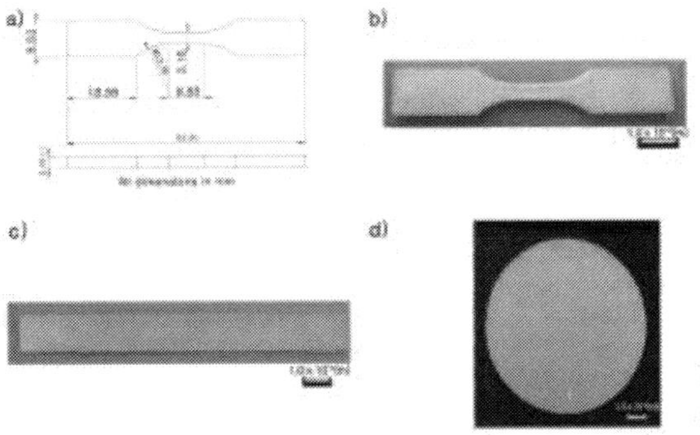

Figure 1. (a) Tensile Testing Standard Specimen specifications for ASTM DS 638 type V, (b) 3D printed sample used for tensile testing, (c) ASTM 790 bending test sample (d) pin-on wheel wear test sample.

Table 1 shows the printing temperature and heat treatment temperatures, for which 90 and 150 % of the glass transition temperatures (T_g) reported in the MSDS sheet from the suppliers. The samples were kept in the furnace for 60 minutes; then the furnace was turned off until room temperature was reached (around 60 minutes). For the bending and wear tests, the optimum printing and heat treatment temperatures were chosen.

Table 1. Printing temperatures and heat treatment temperatures used for PC, PMMA, and PEEK.

Materials	Printing Temp.	Post-Heat Treatment Temp.
PC	232 °C, 255 °C, 280 °C	132 °C, 220 °C
PMMA	232 °C, 255 °C, 280 °C	97 °C, 162 °C
PEEK	345 °C, 355 °C	128 °C, 215 °C

3 RESULTS AND DISCUSSION

Shore Hardness values were obtained from the 3D printed samples before and after heat treatment. The hardness of the printed samples increased after heat treatment, but it mainly decreased when the annealing temperature was at 150 % of T_g.

The tensile strength values for PC, PMMA, and PEEK are 65, 70, and 96 MPa respectively as seen in Figure 2. And also, it shows the tensile results before and after heat treatment. Heat treatment of PC at 90 % of Tg resulted in an increase in the strength of around 10 MPa at all printing temperatures. The heat treatment at 150 % of Tg made the samples to become very brittle, showing a sharp decrease in UTS around 80 MPa to about 40 MPa after the heat treatment.

For PMMA, the results also showed a slight increase in strength for the heat treatment at 90 % of T_g, from about 60 to 80 MPa, and a sharp decrease of strength and ductility when heat treated at 150 % of T_g. Both PC and PMMA showed a reduction in Young's modulus when heat treated at 150% of T_g, as can be observed by the decrease in the slope in all cases. PEEK demonstrated a minimal change in stress after heat treatment.

As heat treatment was increased to 150 % of T_g, the change in Young's modulus was minimum, although the UTS showed a decrease.

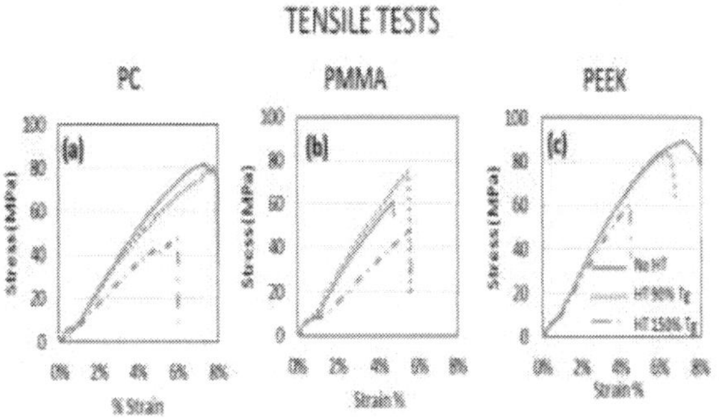

Figure 2. Tensile test results of (a) PC, (b) PMMA and (c) PEEK before and after heat treatment.

Heat treatments at 150 % of T_g, resulted in deformation and change in sample size dimensions. For this reason, it was decided to conduct the bending and wear testing for heat treatments only at 90% of T_g.

Bending tests results are depicted in Figure 3(a) PC, (b) PMMA and (c) PEEK. These tests did not show much improvement in the strength of any of the tested polymers. There was a change in the maximum load of less than 10 %, in all three polymers. The most significant difference between them is the fact that PMMA showed a more brittle behavior, but results were very similar to the ones obtained in tensile tests, while PC and PEEK exhibited distinct behaviors by tensile and bending test. Previous studies have also suggested the results of tensile and bending varies with the size of the specimen, as well as the strain rate. The results indicated that it would be essential to test depending on the future applications of the material.

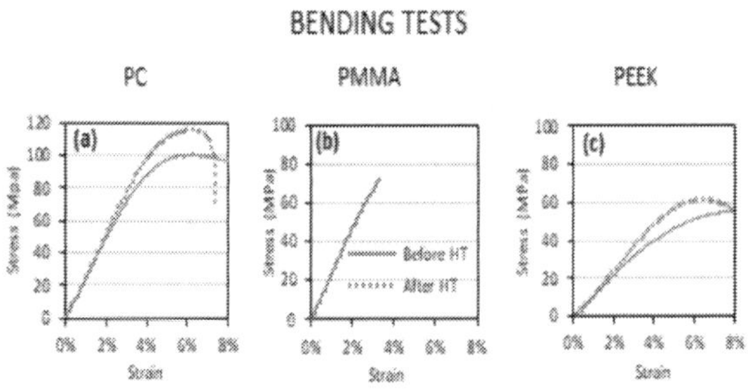

Figure 3. Bending test results of (a) PC, (b) PMMA and (c) PEEK before and after heat treatment.

Wear testing was performed on circular samples measuring 4.0×10^{-2} m in diameter. Figure 4 shows the mean Friction Coefficient values, inset in the graph depicts SEM image of wear test zone showing the worn line.

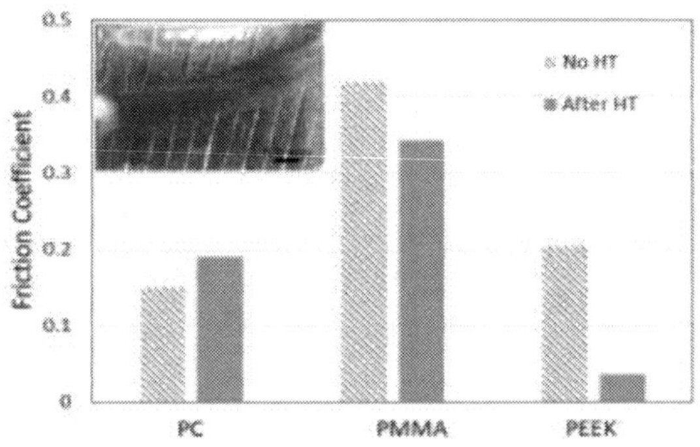

Figure 4. Mean Friction Coefficient values obtained for circular samples of the three polymers used. Inset showing SEM representative image of wear test area on a PMMA sample without heat treatment

For the further application of PMMA and its tribological behavior, wear test was implemented as shown in Figure 5. It was tested as constant the normal force of the pin (5 N), the pin diameter (6.0×10^{-3} m), the wear radius track (1.0×10^{-2} m), the time that the test took place (20 minutes), and the rotation velocity of the disk (30 rpm). Based on the results obtained, Figure 5 also shows a slight decrease in the friction coefficient after heat treatment, which suggests that the surface obtained after heat treatment is more smother in comparison to the one obtained before heat treatment. These results demonstrate that roughness obtained during the 3D printing process was bigger if heat treatment is not performed, which is favorable for the biocompatibility.

Although it is shown that the mechanical properties increased after heat treatment (Figs. 2 and 3), it is important to mention that the roughness of the 3D printed piece decreased which cannot be favorable for the initial cell survival rate and cell attachment, which can lead to a rejection of the piece by the body. The SEM images

(Figure 5), shows the road formed by the wear test, before (a and b) and after (c and d) heat treatment.

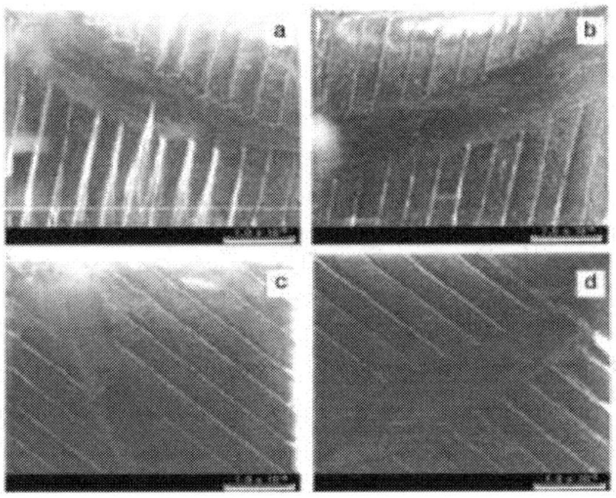

Figure 5 SEM images of the 3D printed PMMA disc after the wear test. (a) & (b) before Heat Treatment, (c) & (d) after Heat Treatment

The SEM images Figure 5 a) and b) shows an irregular and deeper road compared to that of in the c) and d) images. Also, the SEM images in Figure 5 c) and d) show a smoother surface compared to the that in Figure 5 a) and b) images. The comparison shows how the heat treatment influenced the surface properties and the abrasion resistance of the 3D printed material.

Figure 6 shows the relation between the friction coefficient versus time for PMMA samples. In the beginning, a constant and rapid increment of static friction with a slope of 0.0092 is shown with a duration of nearly 100 seconds.

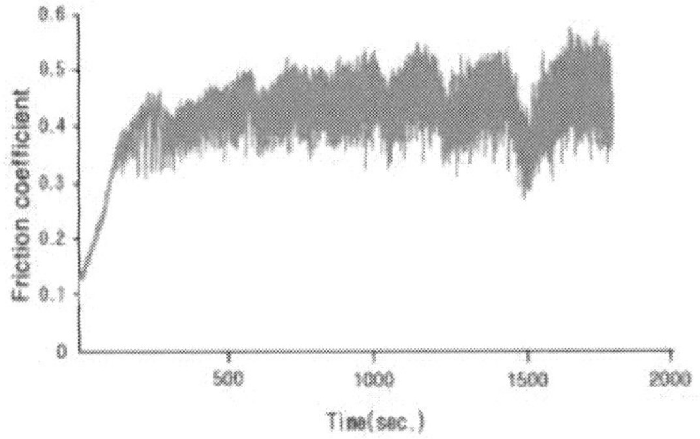

Figure 6 Wear test before heat treatment for PMMA

Figure 7 (sample after heat treatment) shows a slow increment compared to the sample before heat treatment with a slope of 0.0015 and a duration of nearly 700 seconds

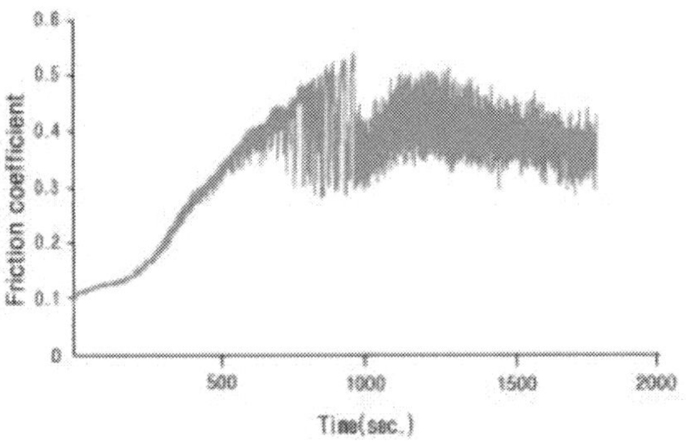

Figure 7. Wear test After Heat Treatment for PMMA

The force required to remove material during the test is less after- than before- heat treatment, which supports the hypothesis that the roughness after heat treatment decreases. The roughness of the surface and the porosity of the sample can lead to the optimal biocompatibility of the cells and a rapid recovery process from the surgery. The effect that roughness has on the surface made by the 3D printing process can be observed in Figure 8. The surface roughness data was obtained from point a to point b in Figure 8 and which is 1.47, where 1 represent a smoother surface and more than one represents a rough surface.

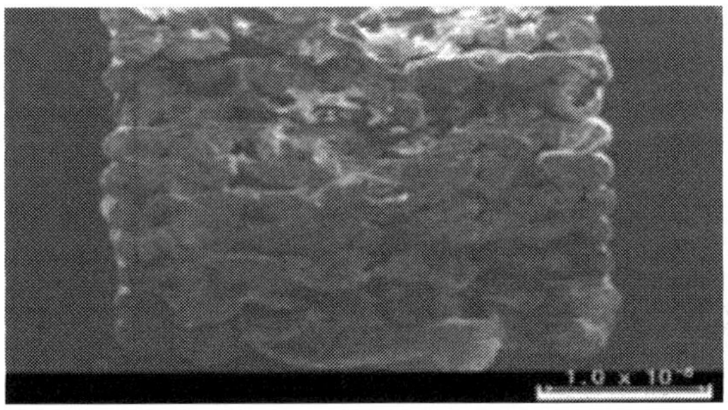

Figure 8 SEM image of the cross-sectional area of PMMA Tensile Testing Standard "ASTM DS 638 type V"

4 CONCLUSIONS

An increase in strength was observed in the polymers used after heat treatment than 3D Printed as-is (before). The increase in strength was due to the recrystallization that the materials undergo during heat treatment. The 3D printing technique deposits fused filament layer by layer, as it solidifies, it undergoes recrystallization in which the molecules are rearranged in an ordered position. However, it is critical not to exceed the Tg temperature of the materials for heat treatment because it results in a decrease in strength. Even though PEEK did not show a change in dimensions as large as the other polymers; PC and PMMA, no benefits were observed in subjecting the polymers to a heat treatment above their T_g

ACKNOWLEDGEMENTS

This study was supported by UTEP-SKU Program: 226-9002-33A "Master Program Using Drug Delivery System (DDS) for Pharmacy Training"

REFERENCES

1. J Hoffman, S Hwang, A Ortega, NS Kim, K Moon., The standardization of printable materials and direct writing systems, Journal of Electronic Packaging **135** (1), 01106 (2013)
2. ASTM D638-14 Standard Test Method for Tensile Properties of Plastics, ASTM International, West Conshohocken, PA (2014), https://doi.org/10.1520/D0638-14.
3. A. A. Aly, Heat Treatment of Polymers: A Review, Int. J. Mater. Chem. Phys, **1** (2), 132–140, (2015)
4. ASTM D790 Standard Test Methods for Flexural Properties of Unreinforced and Reinforced Plastics and Electrical Insulating Materials. [Online]. Available: https://compass.astm.org/EDIT/html_annot.cgi?D790+17. [Accessed: 18-Apr-2018].
5. F. P. Bowden and Tabor D., *Friction and lubrication in solids* (1964).
6. K. Tanaka, *Kinetic friction and dynamic elastic contact behavior of polymers* (1984).
7. C. Weckle, Polycarbonate Remains Proven and Preferred for Medical Applications, (Tech Briefs :: Medical Desi Medicine, Second Edi. 2011. gn Briefs., MEdical Design Briefs, (2013. [Online]. Available: https://www.medicaldesignbriefs.com/component/content/article/mdb/features/17693. [Accessed: 31-Jan-2018].
8. D. G. Powell., Medical Applications of Polycarbonate (MDDI Online, Medical Device and Diagnostic Ind., (1998). [Online]. Available: https://www.mddionline.com/medical-applications-polycarbonate. [Accessed: 31-Jan-2018].
9. K. J. Mark HF, Bikales NM, Overberger CG, Menges G., PEEK Biomaterials in Trauma, Orthopedic, and Spinal Implants. 313–320 (1988).
10. S. M. Kurtz and J. N. Devine, PEEK biomaterials in trauma, orthopedic, and spinal implants (Biomaterials, **28** (32), 4845–69 (2007).
11. J. Black, *Biological performance of materials : fundamentals of biocompatibility* CRC Taylor & Francis (2005).
12. J. Black. G. Hastings. *Handbook of biomaterials properties* (1998).
13. KN Han, NS Kim., Challenges and opportunities in direct write technology using nano-metal particles, KONA Powder and Particle Journal, **27**, 73-83 (2009)
14. AK Amert, DH Oh, NS Kim., A simulation and experimental study on packing of nanoinks to attain better conductivity (Journal of Applied Physics, **108** (10) 2010)
15. S Hong, N Kim., Synthesis of 3D Printable Cu–Ag Core–Shell Materials: Kinetics of CuO Film Removal (Journal of Electronic Materials, **44** (3), 823-830 (2015)
16. S.E Najera, M. Michel, and N.Kim.,3D Printed PLA/PCL/TiO2Composite for Bone Replacement and Grafting, MRS Advances, **3** (40), 2373-2378 (2018)
17. S. Ramakrishna, J. Mayer, E. Wintermantel, and K. W. Leong., Biomedical applications of polymer-composite materials: a review, Compos. Sci. Technol, **61** (9), 1189–1224 (2001).
18. J. R. C. Dizon, A. H. Espera, Q. Chen, and R. C. Advincula., Mechanical characterization of 3D-printed polymers, ddit. Manuf, **20**, 44–67 (2018)
19. NS Kim, KH Han, KH Church., Direct Writing Technology for 21st Century Industries—Focus on Micro-Dispensing Deposition Write Technology, Transactions of the KSMTE Spring Conference 511-515 (2007)
20. Kim, N.S., Eo, J.S., & Cho, D. Optimization of Piston Type Extrusion (PTE) Techniques for 3D Printed Food. Journal of Food Engineering, 235(10), 41-49 (2018).
21. KL Mendoza, A Ortega, NS Kim, Optimization of UV LED-Curable Ink for Reverse-Offset Roll-to-Plate (RO-R2P) Printing, J. of Elect. Mat. **44** (3), 784-791 (2015)
22. J. Eo, B. Cepeda, J. Kim, and NP Kim, A New Paradigm of Pharmaceutical Drug Delivery Systems (DDS): Challenges for Space, Time, and Shapes, Innovations in pharmacy, 9(3), 11-24 (2018)
23. Kim, N., Eo, J., &Shin, J., Three-Dimensional Printed Drug Delivery System and Method, US Patent 62/745139 (2018)

Syntactic core composite sandwich (SCCS) with thermoplastic foam and expanded glass granulate core

Alexander Hackert[1,a], Patrick Schaarschmidt[2], Tristan Timmel[2] and Lothar Kroll[1]

[1] *Institute of Lightweight Structures, Technische Universität Chemnitz, 09107 Chemnitz, Germany*
[2] *Cluster of Excellence MERGE, Technische Universität Chemnitz, 09107 Chemnitz, Germany*

Abstract. Ultra light, highly porous materials arranged as a sandwich structure with thermoplastic fiber reinforced composites (TP-FRC) as a cover layer have very good physical and mechanical properties. An essential prerequisite for the production of hybrid material composites is therefore the provision of material compatible and cost effective manufacturing processes. The lightweight potential of such a syntactic core composite sandwich (SCCS) with thermoplastic and expanded glass granules was examined with the aim of a mass production. After developing a manufacturing strategy, various samples are prepared and their mechanical properties are evaluated by performing bending tests and examination of micrographs.

1 Introduction

Increasing environmental and economic demands in mechanical engineering require specific material combinations and adjusted manufacturing processes. In mechanical engineering, the greatest potential for lightweight construction lies in the reduction of moving masses. Targeted lightweight construction allows the use of smaller drives and the design for lower forces. Thanks to lightweight composite materials, higher machining accuracy, speed, longer service life and lower material wear can be achieved. So it is possible in addition to the targeted adjustment of stiffness and strength, thoughtfully introduced residual stresses, to exploit the anisotropic strength behaviour and apply the special thermal properties. Due to the achievable synergy effects, especially in automotive lightweight design, multi-material assemblies are increasingly applied. Therefore, carbon fibres reinforced plastics (CFRP) arranged with new isotropic syntactic foam structures to a Multi-material Sandwich Composite (MSC) form the basis for novel weight-optimized as well as cost-effective applications. This leads to high specific stiffness and strength structures combined with excellent energy absorption capacity at high damage tolerances as required in a variety of technical applications.

2 Manufacturing strategy

A particular challenge is the processing of macroscopic expanded glass granules with grain sizes over 2 mm in a thermoplastic manufacturing process. The expanded glass (**Figure 1**) comes in the form of spherical granules with a grain density of 320 kg/m³ and mean grain strength of 1.4 N/mm². In addition, conventional PA6 granules in different degrees of filling are used for the thermoplastic matrix.

Figure 1. Expanded glass granules (left), conventional PA6 granules (right)

Due to the grain size, the processing in an injection moulding process is excluded. The expanded glass granules would be destroyed in the system and the high property potential would no longer be usable. Alternatively, the thermoplastic matrix material can be melted in a mould cavity and the expanded glass embedded. Therefore, a dipping edge tool with solid steel base plates that transfer heat evenly is used (**Figure 2**) in a heated press plant, which has two separately adjustable heating plates, which can each be heated up to 300°C. The heating rates of the two plates are each a maximum of 15 K/min. With the cooling system connected in the test field can also be cooled at about 15 K/min.

[a] Corresponding author: alexander.hackert@mb.tu-chemnitz.de

 https://doi.org/10.1051/matecconf/201926402002

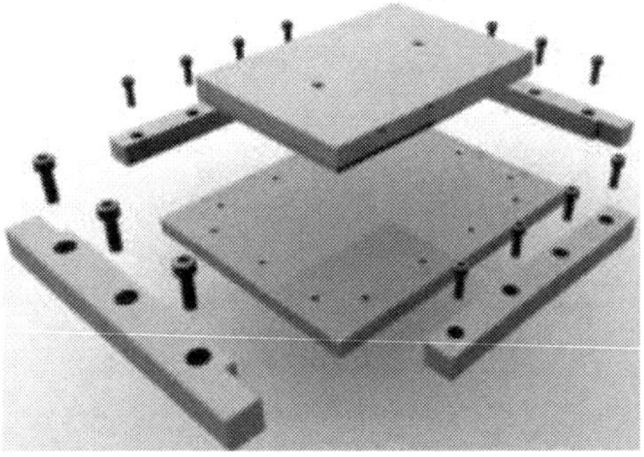

Figure 2. Thermal pressing tool in an exploded view

For the cover layers of the symmetric core composite unidirectional, carbon fibre reinforced, thermoplastic Prepregs are used. These thermoplastic semi-finished products pre-impregnated with PA6 were used in various applications, such as the production of organic sheets or core composites with aluminium foam in many thermoplastic manufacturing processes.

3 Production process

Uniform samples for bending tests and micrograph analyzes were performed for the investigation of the material properties of the manufactured material composite. The production can be divided into three steps: the preparation, the pressing of the individual layers and the remoulding of the sample plate.

3.1 Preparation

First, the unidirectional Prepregs with a PA6 matrix and a fibres volume fraction of 60% are cut to the appropriate shape with approx. 1mm of space per side, so that the fibres in the tool are not compressed. For the samples with fibreglass top layer, three layers per side are used, for the carbon fibre top layers six layers each. In the tool, the core composite is built up in its individual layers (**Figure 3**). The bottom layer is formed by the Prepregs with fibre orientation along the x-axis, followed by a layer of 100 ml or 85 g of PA6 granules. The middle layer is 85 ml or 20 g of expanded glass granules. After that come again a layer PA6 and the upper cover layer.

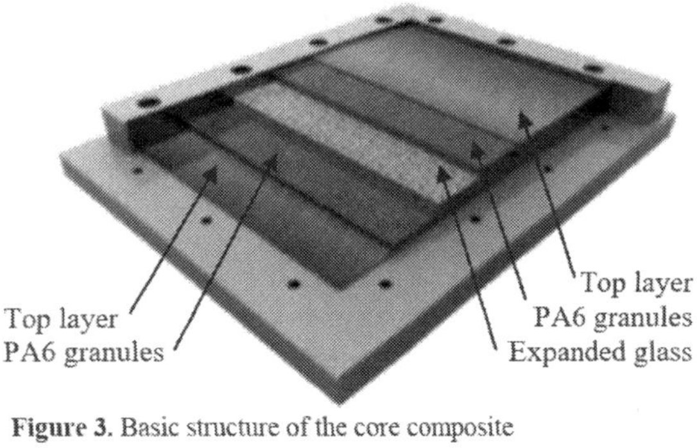

Figure 3. Basic structure of the core composite

In this composition, the core consists of 70% polyamide and 30% expanded glass granules. It should be noted that it is the volume fraction in granular form. The bulk density of unreinforced and reinforced PA6 differs only minimally and is negligible due to measurement errors and differences in the individual grains. The three-layered core structure allows the expanded glass granules to be completely enclosed by the plastic matrix. For two samples, each time with glass and carbon fibre as the topcoat, it is examined whether the matrix eventually flows out of the topcoats into the core when the granules for the core are mixed (**Figure 4**).

Figure 4. Mixed granules

As a further variant, samples without cover layers are made in order to determine the material properties of the expanded glass granules in conjunction with the PA6. For this, 150 ml or 127.5 g PA6 are used as the first layer, after which 128 ml or 29 g expanded glass granules, as well as a further layer PA6. Likewise, samples are made whose cores are composed of a blend ratio of 50% by volume of PA6 and 50% by volume of expanded glass granules, and consist entirely of reinforced PA6. For comparison with currently available core materials, samples are also made with aluminium foam.

3.2 The pressing process

The production cycle is set on the laboratory press without pressure, because the expanded glass granules are not crushed on the one hand and the second not to be pushed too close to the outer layers. Due to the thermal expansion of the tool and of the plastic, a pressure of approximately 10 N/cm^2 arises after the heating phase.

The cycle time is largely limited by the system-side heating and cooling rate of 15 K/min (**Figure 5**).

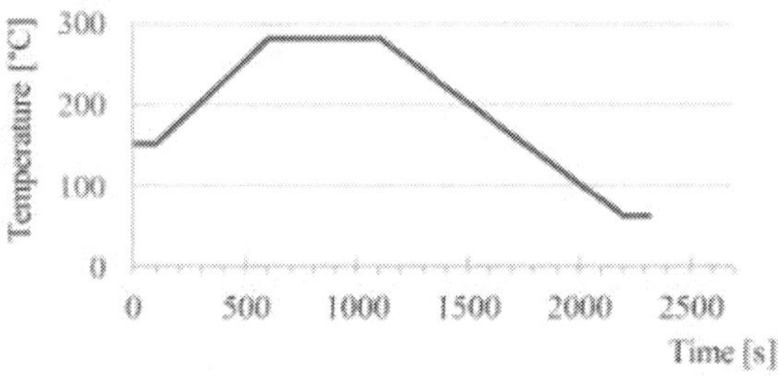

Figure 5. Optimized production cycle

For the first one hundred seconds, the hotplates are held constant at 150°C to bring the tool up to temperature. It is then heated to 280°C and held this temperature for 500 s. Afterwards it is cooled to 60°C. By means of empirical experiments, the machine boundary conditions and the cycle parameters resulting there from during the melting phase are optimized. Overall, there is a cycle time of about 39 minutes.

3.3 Result

During the production of varying samples, a direct influence of the starting materials and their composition as well as their mixing ratio and the mixing strategy could be demonstrated. The unreinforced polyamide sank and the expanded glass granulates floated on the melt. It formed very large air pockets. These have a great influence on the material properties.

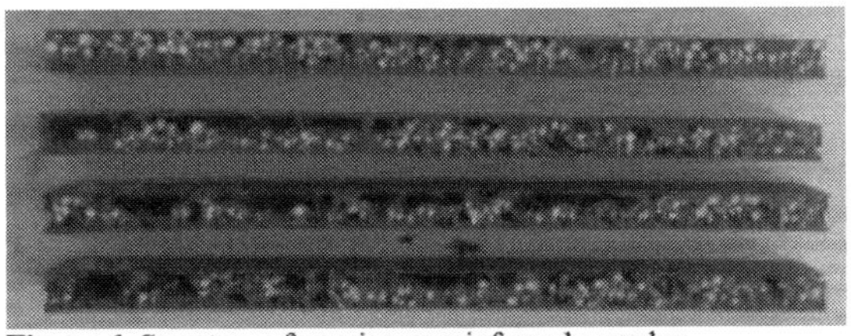

Figure 6. Structure of varying unreinforced samples

In contrast, the reinforced polyamide forms a small-pored, foam-like structure. The fibre reinforcement acts to support and the expanded glass granules are uniformly enclosed by the plastic. Depended on the blend ratios of the reinforced PA6 granulate a varying foam can be adjusted (**Figure 7**). Based on this knowledge, the samples are made for the investigation and optimization of the mechanical properties with fibre reinforced polyamide. A sample was made with a blend of 50% expanded glass granules and 50% glass fiber reinforced polyamide. The specific flexural stiffness in this test is at a level with the 30/70 blend. The higher proportion of expanded glass granules allows a high lightweight construction potential of the core composite and further investigations with different mixing ratios are to be carried out.

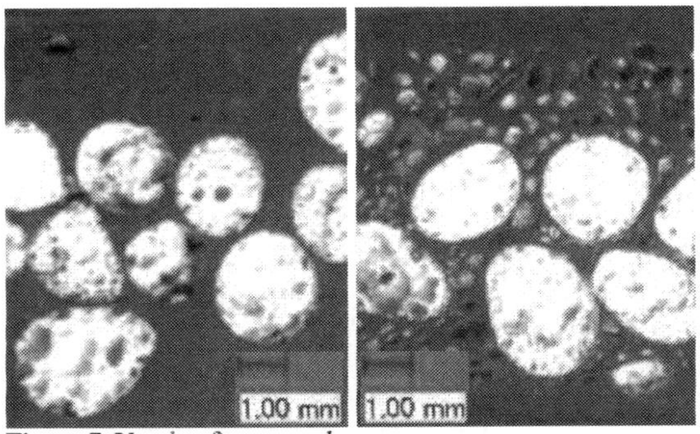

Figure 7. Varying foam samples

A core composite with thermoplastic expanded glass granulate core offers versatile possibilities. Due to the low density of expanded glass granules, it has a high level of lightweight construction potential. Due to the many individual components, there are a variety of combination options, each having different properties. These influencing factors require a closer examination.

4 Examination of mechanical properties

There are a total of 17 manufactured plates for 87 samples with the different variant diversity and an unreinforced aluminium foam plate available for the investigation of which specific flexural modulus is determined by means of 4-point bending tests. In addition, 40 cuttings for micrographs are used to analyze the macroscopic as well as the microscopic structure.

4.1 Bending tests

The tests are carried out in accordance with DIN 53293 for the testing of core composites. With the 4-point bending test characteristic values of the core composite are to be determined under bending stress. In particular, the bending moment at failure, the tensile and compressive stress in the outer layers, the specific flexural modulus and the specific bending stiffness.
During the bending test, the deflection and the pressing force are recorded. By means of the secant method, by two points in the linear range of the measured values ("Hooke's straight line"), the deflection can be used to calculate the flexural modulus E, where ΔF is the change in the pressing force and Δf is the change in the deflection between these points.

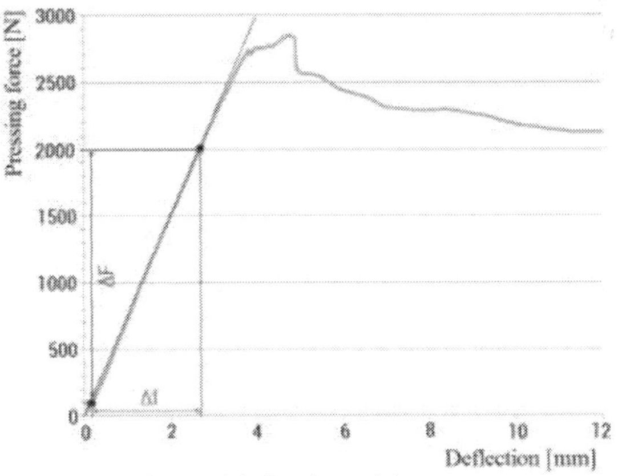

Figure 8. Hooke's straight line (example)

These specific values can now be divided according to the respective core material displayed and analyzed statistically. Here, the region of the second and third as well as the median Quanitl display. The respective largest and smallest measurement is acquired using the error representation. **Figure 9** compares the specific bending stiffnes of variants with carbon fibre top layer.

https://doi.org/10.1051/matecconf/201926402002

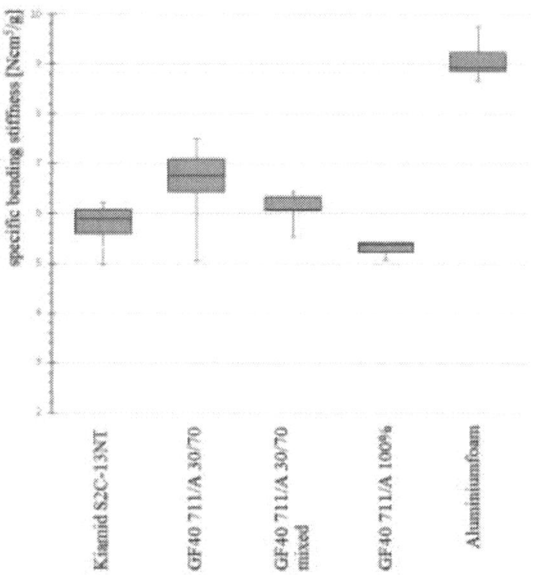

Figure 9. Specific bending stiffness

It turns out that it makes a big difference whether the plastic granulate and the expanded glass granules are applied in layers or mixed directly. However, further investigations are necessary for this. The illustration also shows that the specific bending stiffness of the core without expanded glass granules is significantly lower than all variants with reinforced expanded glass-plastic core. The core values of the mixed granules reach the best values. The unreinforced polymer matrix again turns out to be insufficient. Although the determination of the specific bending stiffness of the samples without cover layer by the 4-point bending test is not universally valid, the individual core materials can be compared with each other. It should be noted that the specific bending strength of the aluminium foam 2.7 times as large as the module of the unreinforced PA6. The reinforced polyamide is approximately in the middle.

4.2 Micrographs

In addition to the flexural stiffness tests, many essential properties can also be determined by analyzing micrographs.

For this purpose, the sample plate is cut at a suitable location and the individual blanks embedded in a red-coloured epoxy resin, so that you get a quasi-flat surface. Afterwards, the sample is ground with ever finer sandpaper until the cut surface is exposed. The last step is to polish the surface.

When grinding and polishing, glass particles are always removed from the blown glass grains, so that it is not possible to obtain a completely scratch-free surface. These are visible in the micrographs, but interfere with the analysis only insignificantly. In order to save embedding always several samples are housed in a specific form as shown in **Figure 10**. The micrographs are viewed under a digital reflected light microscope, with a maximum magnification of 3000x, which allows a direct analysis of the micrographs after recording.

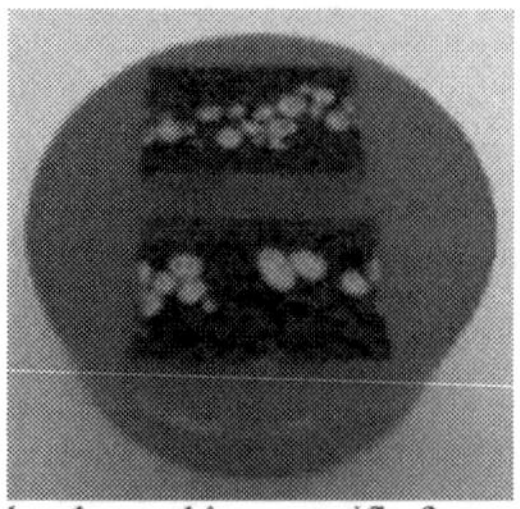

Figure 10. Samples, housed in a specific form

In the micrographs, the following material properties can be examined, which are described in more detail in the following chapter:

- Phase differences and their distribution
- Pore size and distribution

4.2.1 Phase differences

In the micrographs, the individual phases or materials can be clearly identified. These are composed of the polyamide matrix, the reinforcing fibres, the expanded glass granules and air inclusions.

Accordingly, the distribution of the expanded glass granules can be analyzed in the macroscopic range. When in samples with unreinforced polyamide as the core material is a "floating" of the expanded glass granules can be seen, with a large part of the granules collected in the upper part of the samples. Also, it can be seen very well on the basis of this micro section, the triggering of the glass particles.

In the samples with reinforced polyamide as the core matrix, a more even distribution of core matrix, expanded glass granules and air inclusions occurs, which explains the lower variability of the material properties in the bending test. In all micrographs, it should be noted positively that the expanded glass granulate is uniformly enclosed by the matrix. Even between closely adjacent granules, a separating layer of matrix material is detectable.

At the boundary layer between the matrix and the expanded glass granules, a dark border form, which at first sight might be interpreted as a gap (see **Figure 11**). To study this more thoroughly, a depth analysis was performed. With the measuring system of the reflected-light microscope, a series of images can be created with a small change in the focal length, which are assembled directly into a three-dimensional system in the system.

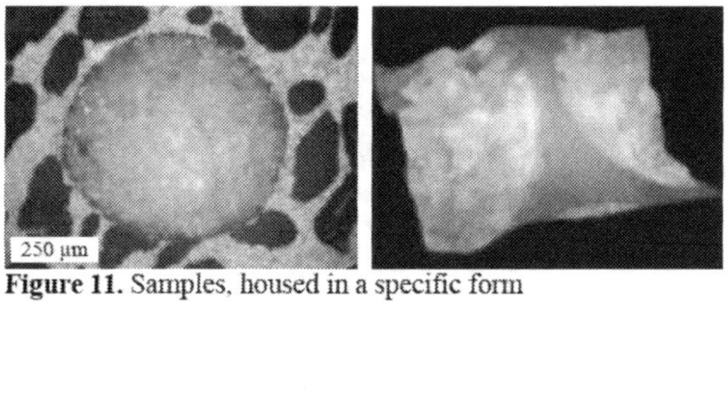

Figure 11. Samples, housed in a specific form

At the transitions from the cover layer to the matrix material, no phase transition is apparent. The glass fibres of the fibre-reinforced polyamide extend directly to the fibres of the outer layers, but the fibres of the outer layers are clearly visible. The missing boundary layer shows that the cover layers are optimally connected to the core.

4.2.2 Pore size and distribution

In the unreinforced polyamide core large air pockets occur, which are distributed irregularly. An analysis is therefore not possible. A closer examination is only possible with the samples with reinforced core, which can also be performed directly with the measuring system of the microscope. Macroscopically, these samples show a uniform distribution of the pores. The evaluation is carried out in areas without expanded glass granulates. On the one hand, the diameter of the pores can be measured directly. On average, over six micrographs and 332 measuring sites, the mean pore diameter is 185 μm at the cutting planes. However, since this is a section through a three-dimensional structure, these values are not very meaningful individually. The average, maximum pore diameter is probably much larger. More meaningful is the comparison of the pore area to the total area. Assuming that the pores are randomly distributed throughout the core and extend evenly in all three dimensions, quantitative statements can be made. For this analysis, the micrograph is decolorized and divided into two areas, black and white. From these areas, the respective area can be determined and compared. This procedure is shown in **Figure 12**.

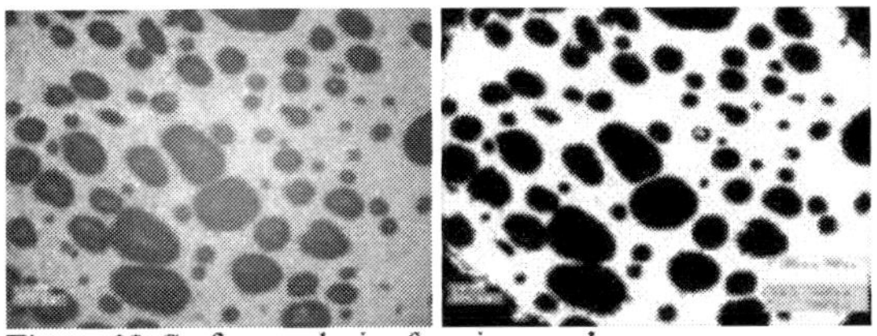

Figure 12. Surface analysis of a micrograph

After analyzing 8 micrographs, the average ratio is 44% matrix and 65% pores. If one deducts the deviations in the border areas one arrives approximately on a relationship of 50% matrix to 50% air inclusions. In addition to the low density of expanded glass granules, this brings another important factor in the weight savings of the core composite.

5 Summary

Compared with core materials currently available in the prior art, the core composite with thermoplastic glass fibre reinforced expanded glass granule core has similar mechanical properties. It should be emphasized positively that the cover layers are firmly bonded to the core, which makes delamination in the boundary region between the cover layer and the core difficult. In 4-point bending, most reinforced core samples fail due to the shear forces in the core. The characteristic of this failure is a 45° inclined to the surface fracture. Occasionally it occurred in the bending samples with unreinforced PA6 to the top layer break at the force application points. This may be due to a higher toughness of the core material. Due to the fibre reinforcement, the core is much stiffer and can dissipate much less energy via plastic deformation. This is especially negative for crash structures.

It should also be noted that expanded glass granules in the vicinity of the cover layers can change the thickness of these. However, it has to be further investigated whether these negatively affect the properties. Since the fibres in the cover layers are continuous fibres, they are only shifted, but not interrupted, and can pass on the tensile forces unhindered. At pressure stress, this deformation can already lead to failure, however, further investigations are necessary because any support effects relatives the buckling of the fibres again.

Due to the large variation of the specific bending stiffness in the samples with reinforced polyamide, further bending tests must be carried out in order to be able to make more precise statements. The individual constituents of the core material were measured by weight. Due to different shapes and granule sizes, however, it will make more sense in the future to carry out the dimensioning by volume. For this a suitable procedure has to be found.

Acknowledgement

These studies were performed as part of a pre-investigation within the Federal Cluster of Excellence EXC 1075 "MERGE Technologies for Multifunctional Lightweight Structures" and supported by the German Research Foundation (DFG) to explore integrative in-mold processes with plastic / metal composites. Financial support is gratefully acknowledged.

References

1. Hackert, A.; Timmel, T.; Drebenstedt, C.; Osiecki, T.; Kroll, L.: *Composite Sandwich with Aluminium Foam core and adhesive bonded Carbon Fibre Reinforced Thermoplastic cover layer*. In: Key Engineering Materials 2017, No.744, pp. 277-281 DOI: 10.4028/www.scientific.net/KEM.744, ISBN-13: 978-3-0357-1116-5
2. Nestler, D. J.: *Beitrag zum Thema Verbund-werkstoffe - Werkstoffverbunde. Status quo und Forschungsansätze*, Habilitation, Technische Universität Chemnitz, 2012.
3. Hackert, A.; Müller, S.; Kroll, L.: *Lightweight Wheel Disc with Carbon Aluminium Foam Sandwich*. Lightweight Design worldwide, March 2017, Volume 10, Issue 1, pp. 6-11, DOI: 10.1007/s41777-017-0016-2, ISSN 2510-2877
4. Osiecki, T.; Gerstenberger, C.; Hackert, A.; Seidlitz, H.; Kroll, L.: *Thermoplastic Fiber Reinforced/Metal-Hybrid Laminates for Structural Lightweight Applications*. 23rd Annual International Conference on Composites/Nano Engineering (ICCE-23). Chengdu (China), 2015.

Study on Thermal Properties and Mechanical Properties of Short-cut Polyimide-Fiber Reinforced Polyphenyl Sulfone Composites

Weizhou Yao[1], Jianan Yao[1], Qing Jiao[2], Yunhai Wei[1], Tianpeng Yu[1], Youhai Yu[1] and Hui Li[1,a]

[1]*School of Materials Science and Engineering, Donghua University, 201600 Shanghai, China*

[2] *Beijing Institute of Remote Sensing Equipment, 100000, Beijing, China*

Abstract. In order to increase the thermal stability and mechanical property of PPSU, two different polyimide (PI) short cut fibers reinforced polyphenyl sulfone (PPSU) composites were prepared by melt extrusion using a three-screw extruder. In addition, the effects of fiber lengths on thermal stability, heat resistance and mechanical properties of the composites was studied. The results indicate that the addition of polyimide chopped fiber can greatly improve the heat resistance of the composites. Comparing with PPSU, with the increasing of fiber content, the heat deformation temperature (HDT) of composites increased from 205 °C to 229 °C, but the addition of polyimide fiber has limited effect on the thermal stability of the composites. Meanwhile, the addition of polyimide chopped fiber can also improve the mechanical properties of the composites. Compared with PPSU, the tensile strength of composites can be increased by 102%, and the bending strength can be raised by 117%.

1 Introduction

With the continuous progress of the society, the requirements for material performance are becoming higher and higher. The excellent properties of composites and modified materials have greatly expanded their application fields. Polysulfone resin is a new type of thermoplastic engineering plastics that has emerged since the 1960s, including polysulfone, polyether sulfone, and polyphenyl sulfone[1-2]. As a typical representative of polysulfone resin, PPSU has excellent mechanical properties, heat resistance, flame retardancy, creep resistance, chemical stability and transparency. It has been widely applied in electronic, automotive, medical and health fields[3].

PI fibers[4], because of its excellent mechanical properties and resistance to radiation ability, excellent thermal stability and good dielectric properties[5-7], are widely used in electricity, microelectronics engineering and space[7-9]. As a kind of high performance organic fiber, PI fiber reinforced composites have attracted increasing attention because of their light weight, high strength and strong toughness in recent years[10]. Because polyimide has a rigid chain molecular structure, its fiber has the excellent mechanical properties, especially in tensile modulus[11]. According to the reports, Russia has developed a high strength polyimide fiber, its tensile strength can reach to 5.8 GPa, modulus up to 285 GPa[12]. Except to have the properties of high strength and high modulus, polyimide fiber also has excellent properties of high temperature resistance and radiation resistance. As an organic fiber, there are many kinds of composite materials made of organic fiber in the world.

Takayanagi[13] used modified Kevlar fibers to enhance LDPE. It was found that when the carboxymethyl volume fraction of the fiber surface was 2.2, the mechanical properties of LDPE increased significantly. Zhao[14] et al. reported that Kevlar fiber and nylon 6 fiber reinforced thermoplastic polypropylene can improve the mechanical properties and heat resistance of composites. Fu[15] investigated on fiber interfacial properties and fiber orientation enhancement effect properties of olefin copolymer composites of Kevlar fiber. Maity[16] and Vaisman[17] studied the UHMWPE fiber reinforced low density and high density polyethylene, and found that the enhancement effect is not obvious.

As far as we all know, it is rarely reported that polyimide fiber as reinforcing materials to prepare composites. In this study, we use two different types, different length of polyimide fiber to reinforce polyphenylene sulfone by extruding and blending. PI fiber reinforcing the PPSU composites can effectively improve the mechanical properties which can apply in harsh environment. Morever, we found that the PI/PPSU composites possessed the best comprehensive properties when the PI fibers content is 30%.

2 Materials and Method

2.1 Material

[a] Corresponding author: lihui@dhu.edu.cn

PPSU powder was purchased from Shandong Haolan special plastic Co., Ltd (China). Short-chopped PI fiber (3 mm, 6 mm) was purchased from Jiangsu SHINO new Mstar Technology Ltd (China). Another short-chopped PI fiber (6 mm) was purchased from Jiangsu AOSHEN new Mstar Technology Ltd (China).

2.2 Preparation and Measurement

PPSU powder and PI fiber were first pre-mixed in a high-speed mixer (HSR-800, Shandong Huanxin Co., Ltd) at 18 rpm for 30 min. The mixture was dried in an oven at 120 °C for 2 h before extrusion. The content of PI fiber added was varied at 10 wt%, 20 wt% and 30 wt% of the blends. The blends were prepared using a three-screw extruder (MEDI-22/40, Guangzhou Putong experimental analysis instrument Co., Ltd.) with a screw speed of 150 rpm at a temperature range of 340~350 °C. The pellets were obtained by a granulator (Magelis, Guangzhou Putong experimental analysis instrument Co., Ltd.) and then shaped by an injection molding machine (TAYU-400, Hangzhou TAYU Machinery Co., Ltd) to produce rectangular and dumbbell-shaped standard specimens. Injection molding conditions were as follows: Injection pressure (90~140 MPa), Injection flow rate (80~99%), Pressure holding pressure (90~140 MPa), barrel temperature (340~350 °C), injection molding temperature (100~120 °C), injection time (10 s) and Pressure holding time (10 s).

The weight loss was measured on a thermogravimetric analyzer TGA 2050 (TA instrument company, USA) under nitrogen and air at a flow rate of 100 mL/min and a heating rate of 10 °C/min from 50 °C to 800 °C. The heat resistance tests were carried out on a Ceast 500 ALOXIDE high temperature thermal deflection temperature/Vicat apparatus with a heating rate of 2 °C/ min and a pressure of 1.82 MPa in accordance with GB/T 1634-2004.

For GB/T 1040.2-2006, the stretch specimens were dumbbell-shaped with dimensions of 75.0 × 12.5 × 2.0 mm^3. For GB/T 9341-2008, the bending specimens had a size of 80 × 10 × 4 mm^3. Tensile test was performed on a Shimadzu AG-1 universal testing machine at room temperature. Three-point bending test was performed on a Shimadzu AG-X (20 kN) universal testing machine at room temperature. Spans for tensile and bending tests were 20 mm and 60 mm, and the crosshead speeds were 5 and 2 mm/min, respectively.

Impact strength was tested on a charpy impact testing machine (HIT-2492, Chengde Jinjian Detection Equipment Co., Ltd) according to GB/T 9341-2000. The impact speed was 2.9 m/s. The data reported were average values of ten samples.

The morphology of the tensile section and the quenching morphology with liquid nitrogen are observed by a scanning electron microscope (SEM-450, FEI, Netherlands), operating at a 15 kV accelerating voltage. The surfaces were all sprayed with gold before scanning.

The DMA Q800 dynamic mechanical analyzer (TA instrument company, USA) was used to measure the composite in air atmosphere. The three point bending mode is used to determine, the temperature range is 50-300 °C, the amplitude is 10 μm and the frequency is 1 Hz.

3 Results and Discussion

3.1. Thermal Stability

TGA and HDT results of different kinds and contents of PI fiber reinforced PPSU composites are presented in Table 1. Both SHINO fiber and AOSHEN fiber, T_{d5} and T_{d10} of the composites were higher than the PPSU. However, with the difference of polyimide fiber content, T_{d5}, T_{d10} and HDT of the composites showed little difference, which indicates that the content of PI fiber has less effect on thermal stability. In addition, different types and different length of PI fiber showed the same heat deformation temperature of composites.

Table 1. Thermal stability of PI fiber reinforced PPSU composites.

Samples	T_{d5}[1]/°C	T_{d10}[1]/°C	HDT[2]/°C
PPSU	560	572	205±2
10%SHINO-3mm	565	576	221±1
20%SHINO-3mm	563	577	220±1
30%SHINO-3mm	564	577	221±1
30%SHINO-6mm	564	576	229±1
30%AOSHEN-6mm	566	576	216±3

[1] 5% weight loss (T_{d5}) and 10% (T_{d10}) weight loss temperatures measured by TGA under air.
[2] Heat deformation temperature.

Figure 1(a) (b) showed that the addition of PI fiber can improve the thermal deformation temperature of the composites, and the thermal deformation temperature is up to 23 °C higher than that of the neat PPSU. This is because PI fiber had a skeleton supporting effect, which hindered the effect of load on material under high temperature. The effect of SHINO fiber on the thermal deformation temperature of composites is greater than that of AOSHEN fiber, this is because the SHINO fiber has a rough surface compared with the AOSHEN fiber, and this lead a better bond properties with the resin matrix. At the same time, the thermal deformation temperature of the composites was also affected by different fiber length. With the same content, the thermal deformation temperature of 6 mm SHINO fiber reinforced composites was higher than 3 mm, which should attribute to the longer length of the fiber, the better supporting effect for composites material, especially under the high temperature.

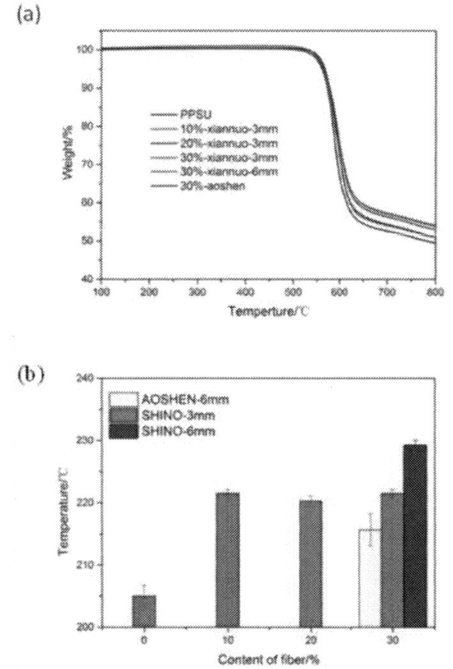

Figure 1. (a) TGA curves of PI fiber reinforced PPSU composites; (b) The bar graph of heat deflection temperature (HDT) of PI fiber reinforced PPSU composites.

3.2 Dynamic Mechanical Analysis

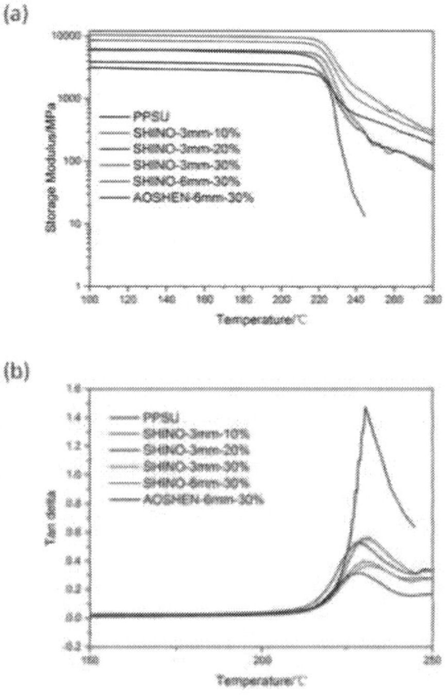

Figure 2. (a) The storage modulus curves of PI fiber reinforced PPSU composites; (b) Tan delta curves of PI fiber reinforced PPSU composites.

The storage modulus and tan delta of the different kinds and contents of the PI fiber reinforced PPSU composites are presented in Figure 2. The storage modulus of PI fiber reinforced composites is between 6.5 GPa~10.7 GPa. Compared with the neat PPSU, storage modulus of the PPSU composites increased greatly, of which 30% SHINO-6mm fiber reinforced PPSU composites were approximately 3.2 times higher than that of the PPSU. In addition, with the increase of fiber content, the storage modulus of composite materials will also increase. Different kinds of fibers also influence on the storage modulus of the composites. With the same content, the storage modulus of SHINO fiber reinforced composites is higher than that of the AOSHEN fiber reinforced composites.

3.3 Mechanical Properties

The mechanical properties of the different kinds and contents of the PI fiber reinforced PPSU composites are listed in Table 2. The tensile strength of the composites increased with the increasing of PI fiber content. The tensile strength of the 30% SHINO fiber reinforced composites increased by about 106% compared with the neat PPSU. With the same content, the tensile strength of the 6mm SHINO fiber reinforced PPSU composite is lower than that of the 3mm SHINO fiber reinforced PPSU composite. The increase of fiber length will lead to fibers entanglement in the matrix resin, so that the tensile properties not be improved. Figure 3 (a) exhibited that the entanglement of 3mm SHINO fiber reinforced PPSU composite is seldom seen, while that of 6mm SHINO fiber reinforced PPSU composite is obvious. AOSHEN fiber can also improve the tensile strength of the composites. Compared with the matrix resin, the tensile strength of 30% AOSHEN fiber reinforced composites was increased by 81%.

Table 2. Thermal stability of PI fiber reinforced PPSU composites.

Samples	Tensile strength (MPa)	Tensile modulus (GPa)	Flexural strength (MPa)	Flexural modulus (GPa)	Impact strength ($kJ\cdot m^{-2}$)
PPSU	69.5	1.0	79.6	2.2	32.2
10%SHINO-3mm	109.1	1.5	118.9	4.2	10.9
20%SHINO-3mm	126.8	1.1	144.7	6.1	11.2
30%SHINO-3mm	140.6	0.6	166.0	7.7	13.5
30%SHINO-6mm	123.7	0.7	172.5	7.2	15.7
30%AOSHEN-6mm	126.3	1.2	94.6	2.8	21.3

The flexural strength of the different kinds and contents of the PI fiber reinforced PPSU composites are presented in the Figure 3 (b). Obviously, the addition of polyimide fibers can improve the flexural strength of the PI fiber reinforced PPSU composites, and the flexural strength of the composites increases with the increase of

 https://doi.org/10.1051/matecconf/201926402003

fiber content. In addition, the flexural strength of the 30% SHINO fiber reinforced composites increased by about 117% compared with the flexural strength of the matrix resin. With the same content, the flexural strength of 6mm 30% SHINO fiber reinforced PPSU composites is higher than that of 3mm SHINO fiber reinforced PPSU composites, and the flexural strength of 6mm SHINO fiber reinforced PPSU composite is lower than 6 mm AOSHEN fiber reinforced PPSU composites.

Figure 3(c) shows the flexural modulus of composites, the addition of polyimide fibers can also improve the flexural modulus of the SHINO fiber reinforced PPSU composites. With the increasing of polyimide fiber content, the flexural modulus of the SHINO fiber reinforced composites increased gradually. However, the addition of fibers does not enhance the flexural modulus of the AOSHEN fiber reinforced PPSU composites. These might be attributed to the interfacial compatibility between the AOSHEN fiber and the matrix resin is not high, which lead to the flexural modulus of the fiber reinforced PPSU composites will no longer increase when reaching a certain content.

As shown in Figure 3(d), the impact strength of fiber reinforced composites through the decline after rising stage, which is attributed to the matrix of the composite was PPSU, its high toughness, high energy absorption per unit area, and joined the polyimide fibers, the toughness and brittleness increased, resulting in impact strength decreases. With the increasing of fiber content, fiber toughness play a role, cause the absorption composite unit area energy increases, the toughness is improved.

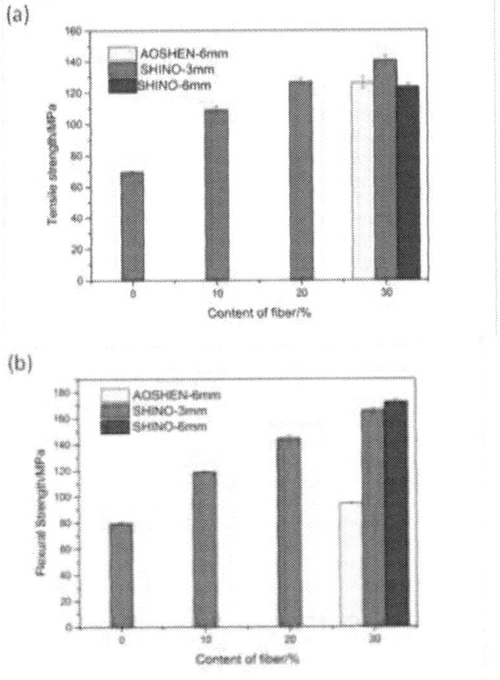

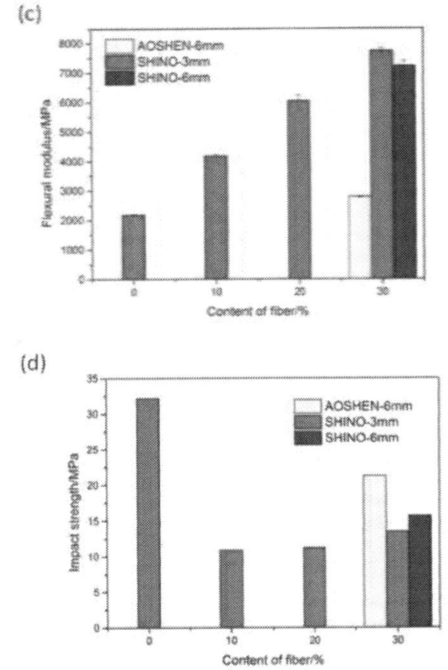

Figure 3. The bar graph of Mechanical properties of PI fiber reinforced PPSU composites. (a) Tensile strength; (b) Bending strength; (c) Bending modulus; (d) Impact strength.

3.4 Scanning Electron Microscope

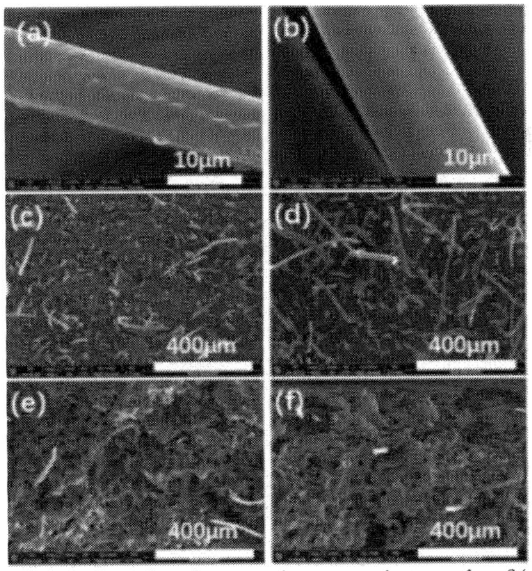

Figure 4. Scanning electron microscope photographs of (a) SHINO fiber; (b) AOSHEN fiber;(c)Tensile section of 3 mm SHINO fiber; (d) Tensisile section of 6mm SHINO fiber composite; (e) Tensile section of 6mm AOSHEN fiber composite. (f) The quenching section of 6 mm AOSHEN fiber composite.

Figure 4 shows the scanning electron microscope photographs of fibers and composites. Two different kinds of fiber with different roughness, the surface of the AOSHEN fiber is more smoothly, but the surface of the SHINO fiber is relatively rough, which leads to a different bonding degree between the two fiber and resin. Figure 4(c), (d), (e), (f) shows that the two kinds of fiber reinforced composites with fewer holes pulled out when the fiber reinforced PPSU composite is stretched, while the holes of the AOSHEN fiber reinforced composite are more than that of the resin, that is because the bonding strength of the SHINO fiber and the resin is higher than the bonding strength of the AOSHEN fiber and the resin. In addition, from these four pictures, we can see that the 3 mm fiber is not curly and entangled in the resin. However, 6 mm fiber reinforced PPSU composites, regardless of the SHINO fiber or AOSHEN fiber, can clearly see the curly and entangled condition.

4 Conclusions

In this article, two kinds of polyimide fiber reinforced PPSU composites were prepared by a parallel three-screw extruder, and subsequently injection molding. The influence of different lengths of PI fiber, different kinds of PI fiber on composites properties were investigated, such as thermal stability and mechanical property. The T_{d5} of two kinds of fiber reinforced composites was in the range of 560 °C~570 °C and T_{d10} was 570 °C~580 °C. The results indicated that polyimide fibers have limited effect on the thermal stability of the composites. However, the addition of two kinds of polyimide fibers can greatly improve the heat resistance of composites. The HDT of PPSU composites containing 30% PI fiber was enhanced from 205 °C to 229 °C comparing with neat PPSU. PI fiber reinforced PPSU composites showed higher tensile strength and flexural strength than neat PPSU, the same as to flexural modulus. Moreover, owing to the SHINO fiber is more rough than AOSHEN fiber, so SHINO fiber with PPSU showed better compatibility than AOSHEN fiber with PPSU, then the SHINO fiber reinforced PPSU composites exhibited better thermal properties and mechanical properties than AOSHEN fiber reinforced PPSU composites.

References

1. Chen,X.; Zhao, J. Q.; Yuan,Y.C.; et al. Synthesis, properties and applications of sulfone polymers. Guangdong Chemistry 2013, 18(40), 66-67.
2. Bian, J.; He, X.F.; Lin, H.L.; Wang, G.; et al. Polyphenylene sulfone/graphene nanocomposites preparation and properties. Journal of Chinese Universities 2015, 29(5), 1230-1237.
3. Wu, J G.; Li, L J.; Guo, H T. Research advance on polysulfone and its composites. China plastics Industry. 2009, 3(7), 9-12.
4. Luong, N O.; Hippi, U.; Korhonen J T.; et al. Enhanced mechanical and electrical properties of polymide film by graphene sheets via in Situ polymerization. Polymer 2011, 52(23); 5237-5242.
5. Niu, H.; Huang, M, Qi, S.; et al. High-performance copolymide fiber contaning quinazolione moiety.: Preparation, Structure and properties[J]. Polymer 2013, 54(6), 1700-1708.
6. Zhang, M Y.; Niu, H Q.; Qi, S L.; et al. Structure evolutions involved in the carbonization of polymide fibers with different chemical constitution. Materials Today Communicaions 2014, 1(1-2), 1-8.
7. Chang,J.; Ge, Q.; Zhang, M.; et al. Effect of pre-imidization on the structures and properties of polymide fibers. RSC Adv 2015, 5(85), 69555-69566.
8. Yin, C.; Dong, J.; Tan, W.; et al. Strain-induced crystallization of polymide moiety. Polymer 2015, 75, 178-186.
9. Chang, J.; Liu, M.; et al. Structures and properties of polymide fibers containing fluorine groups. RSC Adv 2015, 5(87), 71425-71432.
10. Zhang, Q H.; Chen, D J.; Ding, M X. Polymide fiber. Chinese Polymer Bulletin. 2001, 5, 66-72.
11. Liu, X Y; Gu, Y. High performance polymide fibers. New chemical materials. 2005, 5(33), 14-17.
12. Zhang, Q H; Zhang, C H, Chen, D J; Ding, M X. High tech fiber and its application. 2002, 27(5).
13. Takayanagi, M.; Kajiyama, T.; Katayose T. Surface-modified kevlar fiber-reinforced polyethylene and ionomer. Journal of Applied Polymer Science 1982, 27(10), 3903-3917.
14. Zhao, S.; Cheng, L.; Guo, Y.; et al. PA6 and Kevlar fiber reinforced isotactic polypropylene: Structure, mechanical properties and crystallization and melting behavior. Materials &Design 2012, 35,749-753.
15. Fu, S.; Yu, B.; Duan, L.; et al. Combined effect of interfacial strength and fiber orientation on mechanical performance of short Kevlar fiber reinforced olefin block copolymer. Composites Science and Technology 2015, 108, 23-31.
16. Maity, J.; Jacob, C.; Das, C K.; et al. Direct fluorination of UHMWPE fiber and preparation of fluorinated and non-fluorinated fiber composites with LDOE matrix. Polymer Testing 2008, 27(5), 581-590.
17. Vaisman, L.; Gonzalez, M F.; Marom, G. Transcrystallinity in brominated UHMWPE fiber reinforced HDPE comosites: morphology and dielectric properties. Polymer 2003, 44(4), 1229-1235.

The cubic regression model of thermal estimation in the flammability test of the fibrous compound used in bus bodies

Cristian Pérez-Salinas[1,a], Christian Castro[1] and Roberto Valencia[1]

[1]*Universidad Técnica de Ambato, Facultad de Ingeniería Civil y Mecánica, Research group in materials and production, Av. Los Chasquis and Río Payamino, Ecuador*

Abstract. The fire behavior of fiber compounds is a serious concern in automotive applications. The objective of the proposed work is to predict thermal behavior during flammability testing of the composite material of polymer matrix reinforced with fiberglass used in the interior lining of bodies. Different regression models were performed to determine the best fit. This regression analysis determines the existing correlation between the acquired parameters of burn time and temperature versus two types of fibers used for the interior decorative lining. Different regression coefficients were determined and used for the prediction. Through the best fit regression model, thermal behavior of burning during the flammability test is predicted under ISO standard 3795: 1989 and FMVSS 302. The prediction was made for two types of composites, Roof Fiber, FT, and Fiber for laterals, FL. The cubic regression model showed the best prediction fit with a Rq of 0.79 for FT and 0.81 for FL.

1 Introduction

Accidents caused by burning of vehicles has been a constant concern worldwide for the material and human losses that this has entailed. However, the situation differs in different countries. In the US the percentage has decreased substantially but the number is still important [1, 2]. In European countries an increase in numbers has been a constant over time with some decrease in the percentage in recent years [3-5]. Although in Latin America there are no official figures, the report of the written press about the increase in vehicle fires, especially in buses, has increased alarmingly [6-8]. Por lo tanto, la prevención de incendios de vehículos de transporte público es uno los problemas más graves de las estancias gubernamentales [9, 10].

Today, textiles, polymers, composites and various biomaterials are the raw material for the assembly of bodies. A great degree of these materials have an active behavior before the fire. Several investigations have focused on the mechanical characterization and study of technological properties such as their flammability indexes with the intention of improving their behavior in the face of a flame [11-13].

There is a large database available about ignition time, rate of heat released, oxygen index, flame propagation, smoke release and toxicity. At the same time, there have been advances in the development of models for calculating certain properties and estimates of reaction to fire, such as the rate of heat release and the ignition time.

Modeling fire behavior of materials has a great complexity since this phenomenon involves the action of several processes such as chemical, physical, thermal and failure [14, 15]. Given this complexity, the disintegration and study of the different processes allow us to estimate interpreting and predicting the fire behavior of the different materials.

Within the automotive industry, textile, polymer and resinous composites are widely used as structural and decorative materials. Buses use polymer matrix composite materials reinforced with glass fiber to cover the structure of the body interiorly, structurally and decoratively, for example, we have the covering of ceilings, side and rear walls, control panels, drawers, etc. Fire performance is one of the most important factors considered in the design process of products and systems. Therefore, this document describes predictive models of the thermal behavior during the flammability test under ISO 3795: 1989 and FMVSS 302 of glass fiber reinforced polymer matrix composite materials for bus interior application. Allowing to obtain important information for the control and use of materials that benefit the safety of the passengers.

2 Metodology

2.1 Experimental Development

For the study, composite polymer matrix materials reinforced with fiberglass (fiber for roof, fiber for side walls) stratified originally from car body factories where several samples were taken. Some important mechanical

[a] Corresponding author: cf.perez@uta.edu.ec

and technological properties were previously obtained as shown in table 1.

Table 1. Main mechanical and technological properties of fiberglass reinforced polymer matrix composite materials.

Property	Standard	Unit	Value
Tensile strength	ASTM D638	MPa	108
Stretch elongation	ASTM D638	% @ 25 °C	1,9
Flexural strength	ASTM D790	MPa	244
Flammability index	ISO 3795	mm/min	30,58

Specimens were of dimensions 100mm wide, 356mm long and up to 13m thick, as prescribed in ISO 3795: 1989 [16], on an Atlas combustion chamber model HMW under norm FMVSS 302 [17]. A control system (Arduino MEGA 2560 R3 card) was installed to sense the flame temperature inside the chamber through an infrared thermal sensor. The signals were interpreted through the LABVIEW software as shown in the scheme of figure 1.

The specimens were made in shape and dimensions 100mm wide, 356mm long and up to 13mm thick; prescribed in the ISO standard mentioned.

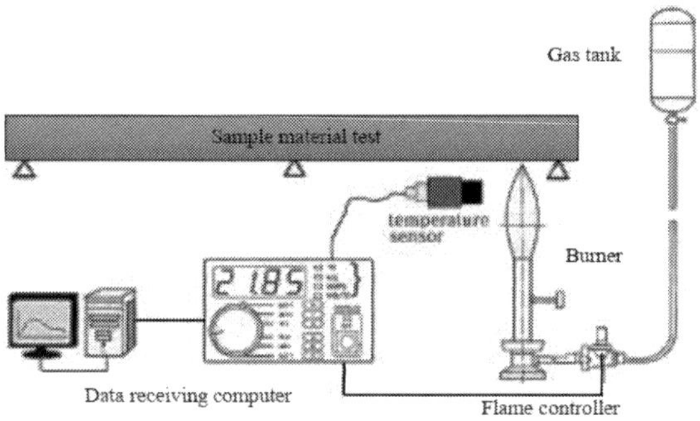

Figure 1. Experimental scheme.

2.2 Statistic Analyses

Six regression models were evaluated to select the best thermal behavior adjustment during the flammability test time. The models evaluated were: Linear, quadratic, cubic, compound, growth and exponential [18].

The selection of the model of best fit provision for each variable was made based on four statistical criteria. An analysis of variance of the regression, coefficients of determination (R2) and significance of the model were made to analyze the goodness of fit of said models. The statistical software "SPSS Estudio" was used to process the data to obtain a thermal behavior model.

3 Results and Discussion

Table 2 shows the results of the measurements obtained by the temperature sensor close to the initial point of the flammability test measured during the average test time in composites (1500 sec).

Due to the prolonged time of resulting flammability test, side fiber (FL) and roof fiber (FT) are considered non-self-extinguishing materials. The standard flame length at the start of the test was 38 mm. When extinguishing flames after the normed time, it was shown that the speed of transmission of the flame or burn of the material propagated constantly until consuming the total length of the test pieces (figure 2).

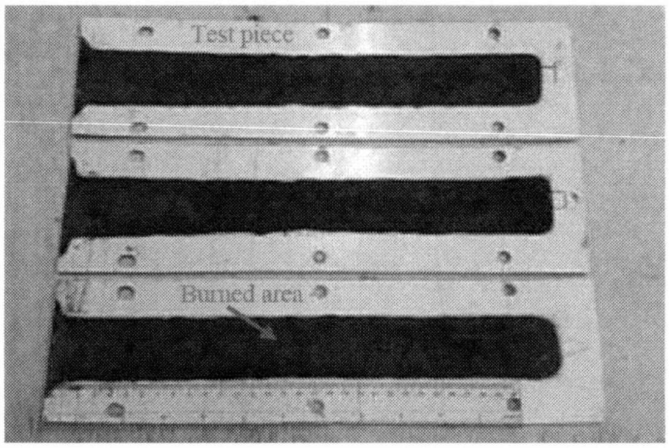

Figure 2. Burned test tubes of flammability test.

Table 2. Temperatures inside the chamber during the time of the flammability test for the FT and FV composites.

Time (seg)	°T sensor FT (°C)	°T sensor FL (°C)	time (seg)	°T sensor FT (°C)	°T sensor FL (°C)
0	45,00	35,75	360	224,17	224,13
1	44,04	35,66	390	226,66	234,96
3	41,47	33,71	420	223,18	240,25
6	40,27	33,06	450	219,02	238,62
9	40,46	32,92	480	208,09	235,54
12	40,92	33,34	510	196,28	224,9
15	41,50	33,45	540	186,79	210,99
18	42,40	33,72	570	177,96	192,11
21	43,46	34,12	600	170,73	179,73
24	44,28	34,7	660	152,84	150,84
27	45,32	35,09	720	128,71	125,56
30	46,32	35,78	780	113,81	105,72
60	55,58	44,4	840	100,53	91,05
90	71,96	56,2	900	88,12	82,33
120	87,04	70,85	960	82,43	77,06
150	102,68	87,7	1020	77,91	73,83
180	121,05	106,7	1080	74,94	70,47
210	146,04	129,1	1230	67,96	66,85
270	188,62	170,69	1380	63,46	62,37
300	205,88	191,75	1530	61,15	59,1
330	219,32	208,66	-	-	-

Once the statistical data treatment was carried out, it was determined that the "cubic model" provided a better adjustment of the thermal behavior during the flammability tests. Figures 3,4,5,6 graphically appreciate the fit of each model with respect to the experimental measurements observed.

Table 3. Summary of the best model adjustment.

Material	R	R square	R square adjusted	Standard error estimate
FT1	0,886	0,785	0,769	31,856
FT2	0,807	0,651	0,636	38,499
FL1	0,900	0,810	0,796	33,256

FL2	0,849	0,720	0,708	33,058

Fiber roof replica 1 (FT1) *Fiber roof replica 1 (FT1)*
Fiber walls replica 1 (FL1) *Fiber walls replica 2 (FL2)*
** The independent variable is time.*

Table 4. ANOVA data analysis for each material.

item	Sum of squares	gl	quadratic mean	F	Sig.
Fiber roof replicate 1 (FT1)					
Regression	151912.09	3	50637.365	49.90	,000
Residue	41605,84	41	1014,777		
Total	193517,93	44			
Fiber roof replicate 2 (FT2)					
Regression	196360,91	3	65453,636	44,160	,000
Residue	105235,02	71	1482,183		
Total	301595,93	74			
Fiber walls replicate 1 (FL1)					
Regression	189040,29	3	63013,432	56,976	,000
Residue	44238,721	40	1105,968		
Total	233279,01	43			
Fiber walls replicate 2 (FL2)					
Regression	199733,12	3	66577,706	60,922	,000
Residue	77591,721	71	1092,841		
Total	277324,84	74			

Table 5. Coefficients of cubic model determined.

item	Non-standardized coefficients		Standardized coefficients	t	Sig.
	B	Standard error	Beta		
Fiber roof replicate 1 (FT1)					
(°T)	0,615	0,054	5,453	11,458	0,00
(°T) ** 2	-0,001	0	-12,783	-10,36	0,00
(°T) ** 3	2,37E-07	0	7,494	9,027	0,00
(Constant)	40,727	8,752		4,654	0,00
Fiber roof replicate 2 (FT2)					
(°T)	0,425	0,066	4,201	6,433	0,00
(°T) ** 2	-0,001	0	-11,974	-7,606	0,00
(°T) ** 3	1,71E-07	0	7,413	7,476	0,00
(Constant)	83,593	15,748		5,308	0,00
Fiber walls replicate 1 (FL1)					
(°T)	0,742	0,06	5,494	12,327	0,00
(°T) ** 2	-0,001	0	-12,404	-10,88	0,00
(°T) ** 3	3,12E-07	0	7,159	9,38	0,00
(Constant)	24,605	9,22		2,669	0,011
Fiber walls replicate 2 (FL2)					
(°T)	0,482	0,057	4,967	8,495	0,00
(°T) ** 2	-0,001	0	-13,446	-9,538	0,00
(°T) ** 3	1,81E-07	0	8,142	9,171	0,00
(Constant)	66,278	13,523		4,901	0,00

Table 3 shows the result of the cubic model of best fit for each material tested. A cubic prediction can be made, since it has an adjustment of the model, greater than 0. 7 of R squared and a relation between the two variables positive average of 0.85 for the two materials tested with their respective replicas (Table 3).

The ANOVA analyzes for all replicas show that the modeling of the cubic regression is appropriate, since the level of significance is zero for all cases (Table 4).

Table 5 shows the coefficients of the cubic model, the constant, the coefficient of the variable, the coefficient of the quadratic variable and the coefficient of the cubic variable. Each of them is significant, that is, they have a value less than 0.05 with a confidence level of 95%. In this way, the four equations or cubic models are formed to make a prediction of the time and temperature variable.

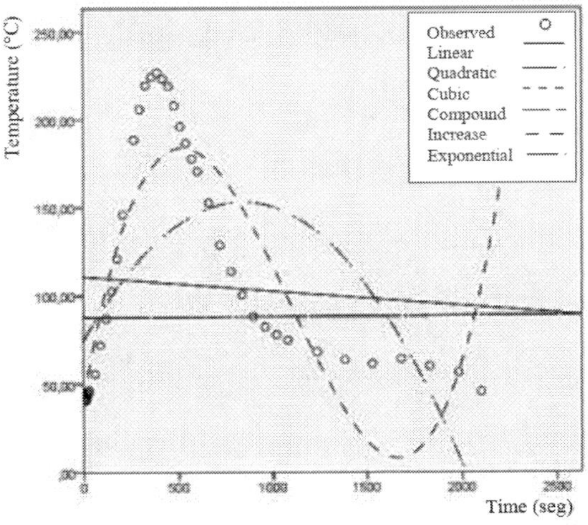

Figure 3. Regression models for thermal behavior of FT1

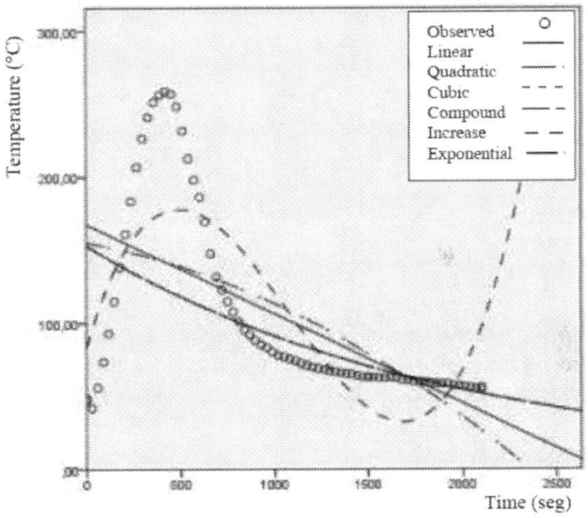

Figure 4. Regression models for thermal behavior of FT2

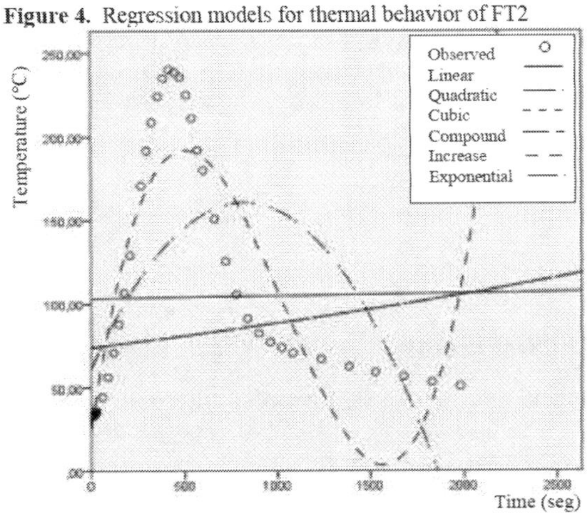

Figure 5. Regression models for thermal behavior of FL1

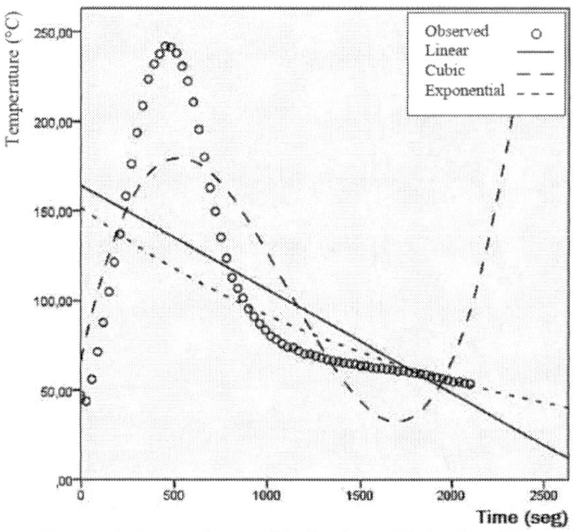

Figure 6. Regression models for thermal behavior of FT1

The obtained equations representing the in-chamber temperature modeling for roof composites (FT) and for side walls (FL) during the initial experimentation and their replications observe in equations 1,2,3 and 4 respectively:

$$MeanFT_1 = 40,73 + 0,62(°T) - 0,001(°T)^2 + 2,37x10^{-7}(°T)^3 \quad (1)$$

$$MeanFT_2 = 83,60 + 0,43(°T) - 0,001(°T)^2 + 1,71x10^{-7}(°T)^3 \quad (2)$$

$$MeanFL_1 = 24,61 + 0,74(°T) - 0,001(°T)^2 + 3,12x10^{-7}(°T)^3 \quad (3)$$

$$MeanFL_2 = 66,28 + 0,48(°T) - 0,001(°T)^2 + 1,81x10^{-7}(°T)^3 \quad (4)$$

Analyzing each of the prescribed equations, it follows that the regression models that best adapt for a statistically reliable prediction are equations 1 and 4. This is because it has a higher Fisher coefficient of determination (F = 49.90 and F = 60.92) for each respective material, which means that the models come from. All the coefficients are statistically significant since they are less than 0.05, with a confidence level of 95%. Consequently, we can predict the thermal behavior of the composite material of polymer matrix reinforced with glass fiber for roof and side walls individually. Starting from the fact that the two applications are made with the same material, but with different number of layers and different thickness of surface treatment (Gelcoat to texturize it), it can be said that the best predictive model is equation 4 because of its higher coefficient value from Fisher (F = 60.92). This is supported by the fact that the results of the thermal behavior inside the chamber have significant differences.

4 Conclusion

Unlike other structural materials such as steel, aluminum alloy and reinforced concrete, the compounds are reactive materials at high temperature due to the polymer matrix that has. Consequently many authors have reported the need to incorporate additives [15, 19].

Regression models have been applied in different fields and are a valuable tool in the study and interpretation of systems with diverse characteristics [10, 20-22].

All the models analyzed, the cubic regression model shows the best fit to the phenomenon studied. Equations 1 and 4 present the best prediction of the thermal behavior for the FT and FL composites respectively, all coefficients being significantly lower than 0.05, with a confidence level of 95%.

The thermal models developed to calculate the temperature distribution inside the combustion chamber under norm ISO 3795: 1989 and FMV 302 in the laminated compounds studied can predict with good precision the behavior of the temperature, according to the statistical results found in the Table 3. This is consistent with what has been mentioned in similar studies that indicate the importance of analyzing parallel phenomena such as physicists and chemists to better understand the behavior of these materials in the face of fire. [23-25]. On the other hand, also the importance of analyzing other materials such as textiles, thermoplastics that are also used in the manufacture of buses and automobiles.

The burning temperature of the materials reaches its peak (238 ° C for FL and 224 ° C for FT) at an average time of 400 seconds from the start of combustion and remains at a temperature higher than 200 ° C during a time of 150 to 200 seconds. The thermal energy deployed in the lateral fiber (FL) is slightly higher than in the roof fiber (FT) because the "Gelcoat" (coating material for surface texture copy) has greater thickness to be mixed with decorative coloring additive in FL, as it is also due to the fact that said fiber has an additional layer of fiberglass with respect to FT.

References

[1] M. Ahrens, "Trends and patterns of US fire loss," *National Fire Protection Association (NFPA) report Google Scholar,* 2017.

[2] M. Ahrens, *US vehicle fire trends and patterns*: National Fire Protection Association, 2010.

[3] B. Reinicke, J. Axelsson, and R. Hammarström, "Bus and coach fires in Sweden and Norway," ed, 2006.

[4] Y. Li and M. Spearpoint, "Analysis of Vehicle Fire Statistics in New Zealand Parking Buildings," *Fire Technology,* vol. 43, pp. 93-106, June 01 2007.

[5] A. Hofmann and S. Dülsen, "Fire safety performance of buses," *Fires in Vehicles-FIVE 2012,* p. 147, 2012.

[6] Periódico, "El pais" publicación 04 de julio de 2018," ed: Colombia.

[7] Periódico, "El Telégrafo", publicación de 04 de septiembre de 2018," ed: Ecuador.

[8] Periódico, "El Heraldo" publicación de 25 de febrero de 2018.," ed: Ecuador.

[9] A. Cordner, M. Mulcahy, and P. Brown, "Chemical regulation on fire: rapid policy advances on flame retardants," *Environmental science & technology,* vol. 47, pp. 7067-7076, 2013.

[10] L. H.-s. Z. Xin-lei and S. Li-xia, "Comprehensive Evaluation and Prediction of Fire Accidents in China

Based on Statistics [J]," *China Safety Science Journal,* vol. 6, p. 010, 2011.

[11] C. Q. Yang, Q. He, R. E. Lyon, and Y. Hu, "Investigation of the flammability of different textile fabrics using micro-scale combustion calorimetry," *Polymer Degradation and Stability,* vol. 95, pp. 108-115, 2010/02/01/ 2010.

[12] D. Gao, R. Li, B. Lv, J. Ma, F. Tian, and J. Zhang, "Flammability, thermal and physical-mechanical properties of cationic polymer/montmorillonite composite on cotton fabric," *Composites Part B: Engineering,* vol. 77, pp. 329-337, 2015/08/01/ 2015.

[13] H. Yang, Q. Fu, X. Cheng, R. K. K. Yuen, and H. Zhang, "Investigation of the Flammability of Different Cables Using Pyrolysis Combustion Flow Calorimeter," *Procedia Engineering,* vol. 62, pp. 778-785, 2013/01/01/ 2013.

[14] A. Mouritz, S. Feih, E. Kandare, Z. Mathys, A. Gibson, P. Des Jardin, *et al.*, "Review of fire structural modelling of polymer composites," *Composites Part A: Applied Science and Manufacturing,* vol. 40, pp. 1800-1814, 2009.

[15] M. Prabhakar, A. U. R. Shah, and J.-I. Song, "A review on the flammability and flame retardant properties of natural fibers and polymer matrix based composites," *Polymer,* vol. 37, p. 40, 2015.

[16] ISO, "ISO 3795: 1989. Road vehicles, and tractors and machinery for agriculture and forestry -- Determination of burning behaviour of interior materials," p. 6.

[17] FMVSS, "FMV 302 Flammability of interior materials," *Federal Motor Vehicles Safety Standards.*

[18] R. B. Darlington and A. F. Hayes, *Regression analysis and linear models: Concepts, applications, and implementation*: Guilford Publications, 2016.

[19] A. Subasinghe, R. Das, and D. Bhattacharyya, "Study of thermal, flammability and mechanical properties of intumescent flame retardant PP/kenaf nanocomposites," *International Journal of Smart and Nano Materials,* vol. 7, pp. 202-220, 2016.

[20] L. Rodríguez, V. Torres, R. O. Martínez, O. Jay, A. C. Noda, and M. Herrera, "Modelos para estimar la dinámica de crecimiento de Pennisetum purpureum vc. Cuba CT-169," *Revista cubana de ciencia agrícola,* vol. 45, 2011.

[21] S. Chithra, S. S. Kumar, K. Chinnaraju, and F. A. Ashmita, "A comparative study on the compressive strength prediction models for High Performance Concrete containing nano silica and copper slag using regression analysis and Artificial Neural Networks," *Construction and Building Materials,* vol. 114, pp. 528-535, 2016.

[22] P. V. Parandekar, A. R. Browning, and O. Prakash, "Modeling the flammability characteristics of polymers using quantitative structure–property relationships (QSPR)," *Polymer Engineering & Science,* vol. 55, pp. 1553-1559, 2015.

[23] F. Gharagheizi, "Prediction of upper flammability limit percent of pure compounds from their molecular structures," *Journal of Hazardous Materials,* vol. 167, pp. 507-510, 2009/08/15/ 2009.

[24] A. Bilal, R. J. T. Lin, and K. Jayaraman, "Optimisation of material compositions for flammability characteristics in rice husk/polyethylene composites," *Journal of Reinforced Plastics and Composites,* vol. 33, pp. 2021-2033, 2014/11/01 2014.

[25] H. Yaghoobi and A. Fereidoon, "Thermal analysis, statistical predicting, and optimization of the flexural properties of natural fiber biocomposites using Box–Behnken experimental design," *Journal of Natural Fibers,* pp. 1-19, 2018.

Experimental study on fiber mortar strength at early age under temperature difference

Yan Zhang [1,a] and Ning Li [2]

[1] *School of civil engineering and architecture, Xi' an University of Science and Technology, Xi' an, Shaanxi 710054, China*
[2] *Institute of Geotechnical Engineering, Xian University of Technology, Xi' an, Shaanxi 710048, China*

Abstract. The condition of high temperature is easy to destroy the concrete structure of tunnels. Many temperature difference conditions are applied to ordinary mortar and fiber mortar specimens, the early strength values of mortar caused by temperature change were measured. The flexural strength change rules of the specimens were analysed. The influence of temperature difference on different specimens, and the cause of the strength difference and failure of the specimens are studied. It provides experimental basis for improving the service life of high temperature tunnel cement concrete structure.

1 Introduction

Temperature difference in high temperature tunnels is a geological phenomenon which can be frequently encountered in engineering. According to the structural characteristics of concrete, the condition of high temperature difference creates the condition for the large number of early cracks in concrete [1-3]. Research shows [4-6], at early age when the strength is low, under the action of early temperature stress and dry shrinkage stress etc., the probability of micro cracks in concrete will increase greatly. At the same time, the early performance of concrete changes significantly, and cracks are generated easily. Early stage concrete is affected obviously by temperature and the hydration will be accelerated [7,8]. Temperature will eventually affect the strength of concrete. When the tensile stress exceeds the tensile strength of the concrete, the structure will crack[9,10]. At present, the formation and expansion of cracks in early concrete structures are difficult to predict.

A number of experimental results showed that the tensile strength of concrete at high temperatures had an increasing influence on its structural bearing capacity [11-14]. In this paper, the tensile strength of sprayed concrete was studied by the flexural strength of the specimen. According to the temperature difference state of the cement concrete spray layer, the tensile strength tests of mortar specimens under different temperature difference conditions were carried out. Therefore, the early mechanical properties of cement mortar under different high temperature differential conditions can be investigated.

2 Experiment scheme

The cement used in the experiment was the Shaanxi Qinling Mountains ordinary portland cement (PO42.5R), all indicators conform to national standard GB50010-2010[15]. The mass ratio of sand to cement was set as1.0, and four sets of mortar specimens were formed: plain mortar (OM), polyester mortar (PEM), polypropylene mortar (PPM) and polyacrylonitrile mortar (PAM). The fiber content of this sample was $0.9kg/m^3$, as recommended by the manufacturer. The specific material mix ratio of the sample is shown in Table 1. According to fluidity test, the sand rate was adjusted to maintain a certain fluidity 45mm. The prism of the specimen was 40mm×40mm× 160mm and form removal after the initial solidification. Temperature difference heating device, the electrothermal ring (diameter 300mm, high 100mm) is shown in Figure 1. The temperature of the heating surface can be controlled by an automatic temperature controller, the internal emptying surface is exposed to indoor air at a temperature of 25 ° C. The temperature of the heating surface of the specimen is controlled at 25° C, 45° C, 60° C, 80° C, 90° C, 100 ° C, respectively. That is to control the temperature difference on both sides at 0° C, 20° C, 35° C, 55° C, 65° C, 75 ° C. Then the molded specimens are maintained to the designated 3d and 7 d age and taken out and the flexural strength were measured separately. When measuring, the direction of temperature difference is consistent with the loading direction.

The test circuit includes electric heating coil, temperature probe, temperature controller, overcurrent protection circuit opener and AC contactor. It also

[a] Yan Zhang: ylozy@126.com

includes DH3816 strain tester which can measure strain and the corresponding computer to collect the strain data of other subsequent tests. Flexural strength is determined by cement mortar flexural strength testing equipment DKZ-5000. When the bearing capacity of the specimen is reduced to 85% of the peak load, it is considered as the specimen failure.

Table 1. The proportioning of the mortar specimens.

Types	Fiber content	Cement: water: sand
OM	0	1:0.52:3
PEM	0.9kg/m^3	
PPM	0.9kg/m^3	
PAM	0.9kg/m^3	

Figure 1. Temperature differences curing device for samples.

3 Results and analysis

The flexural strength curves of mortar specimens with different temperature difference at early age are shown in figures 2 and 3. The flexural strength values of all kinds of mortar at different temperatures are shown in figures 4 and 5.

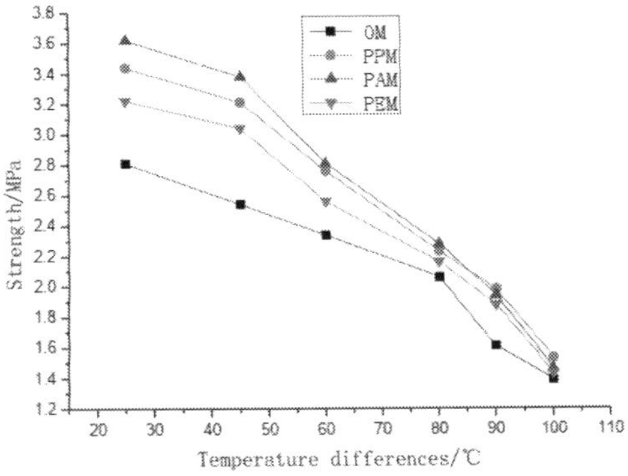

Figure 2. Variation curves of flexural strength of 3 days mortar with temperature differences.

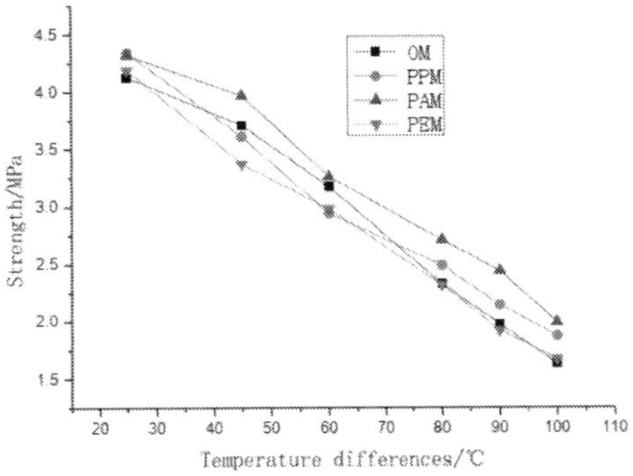

Figure 3. Variation curves of flexural strength of 7 days mortar with temperature differences.

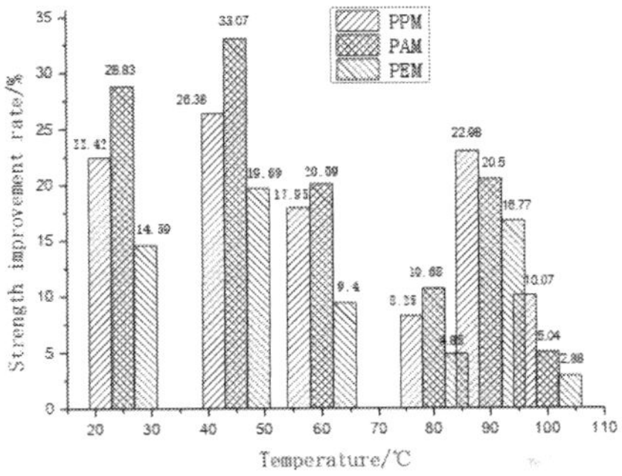

Figure 4. Flexural strength growth rate of 3 days mortar with temperature differences.

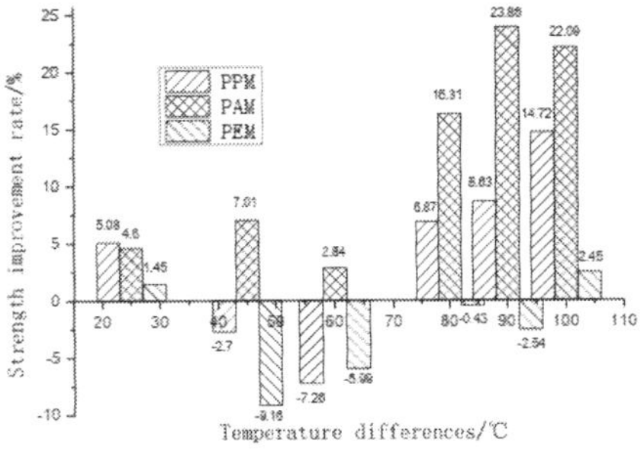

Figure 5. Flexural strength growth rate of 7 days mortar with temperature differences.

It can be seen from figures 2 and 3 that the flexural strength of early-age mortar specimens decreases with the increasing temperature difference. The reduction of fiber mortar flexural strength is smaller than plain mortar under temperature difference. It indicates that the fiber can help to alleviate the impact of temperature difference on the flexural strength of mortar.

It can be seen from figure 4 that the flexural strength of fiber mortar specimens is greater than that of plain mortar at the age of 3d during every temperature range, and the fiber has begun to play a role in the early stage of cement-based mortar. When the temperature difference is 20 ° C, the flexural strength of polyacrylonitrile fiber mortar is greater than the plain cement mortar by 33 %. Compared to plain cement mortar, the flexural strength of polyacrylonitrile fiber mortar increased by 21% and 11% respectively when the temperature difference between 35 ° C to 55 ° C. When the temperature difference is 65 ° C, the flexural strength of polypropylene fiber mortar is higher than the plain mortar by 21%. When the temperature difference is 75 ° C, the flexural strength of polypropylene fiber mortar is higher than the plain mortar by 11%. The effect of polyester fiber on the flexural strength is minimal. It can be seen that the incorporation of fibers will increase the flexural strength of mortars at temperature difference. The smaller the temperature difference, the smaller the effect on flexural strength of mortars degradation, and the less obvious the effect of fibers.

As shown in figure 5, the flexural strength of all mortars at the age of 7 days decreases with the increasing temperature difference, and the reduction rate of the flexural strength is higher than the mortars at the age of 3d. The greatest reduction of the flexural strength is plain mortar. Compared with plain cement mortar, the flexural strength of polyacrylonitrile fiber mortar increased by 5%, 7%, 3%, 7%, 24% and 22%, respectively when the temperature difference is 20 ° C, 35 ° C and 55 ° C, 65 ° C and 75 ° C. Only when the temperature difference is 20 ° C, the flexural strength of polypropylene fiber mortar is higher than that of polyacrylonitrile mortar by 5%, and the flexural strength at other temperatures is not as strong as polyacrylonitrile mortar. When the temperature differences are 20 ° C and 35 ° C, the flexural strength of polypropylene, polyester fiber mortar are less than that of plain cement mortar. It can be seen that polyacrylonitrile fiber mortar in the long age under high temperature conditions has better stability. In general, the greater the temperature difference, the more serious the strength degradation of the specimens. With increasing age, the influence of temperature difference is greater. Temperature difference restricts the growth rate of mortar strength. The greater the temperature difference, the slower the strength growth of mortar specimens in the same time period.

4 Conclusion

The flexural strength of four types of specimens (OM, PEM, PPM, PAM) under temperature differences were investigated. And the experimental results provide a reference for the stability analysis of high temperature tunnels. Based on the relevant results, the following conclusions can be drawn:

1) Through the mortar test under the condition of the temperature difference, the influence rule of flexural strength of mortar at early ages under temperature difference was obtained. Furthermore, the earlier the mortar strength age, the more obviously affected by temperature difference.

2) With the experimental data, the strength change curve of the mortar specimen at different temperature difference is plotted. The results show that the temperature difference has a deterioration effect on the mortar strength. The fibers in mortar can improve the degradation of the flexural strength and play a certain role in alleviating the degradation. The adhesion between fiber and cement matrix prevents the crack from spreading.

3) At the same age, the greater the temperature difference is, the greater the reduction of the strength of the mortar specimen will seriously affect the stability of the cement concrete structure of the high-temperature tunnel. The temperature difference should be solved as far as possible in practical engineering.

Acknowledgements

The study was sponsored by the National Natural Science Foundation of China (Grant No. 51509200), the Nurturing Foundation of Xi'an University of Science and Technology (Grant No. 6310214036) and the Doctoral Foundation of Xi'an University of Science and Technology (Grant No. 6310115020).

References

1. Z.H. Guo, X.D. Shi, Behaviour of reinforced concrete at elevated temperature and its calculation. Beijing: Tsinghua University Press(2002)
2. J.C. Guo, J.T.Yu, Z.D.Lu, Industrial construction. **38**, 9(2008)
3. Y. Li, X. Liu, M. Wu, Const. Build. Mater.**34**, (2017)
4. S. Lepage , M. Baalbaki , É. Dallaire , P.C. AïTcin, Cement, Concrete and Aggregates, **21**,2(1999)
5. X.Y. Jin, Y.Tian, N.G. Jin , Journal Of Building Structures , **31**,6(2010)
6. E. Tazawa. Autogenous shrinkage of concrete. London: Great Britain, St Edmundsbury Press(1998)
7. I. Khan, A.Castel, R.I.Gilbert, ACI Mater. J., **114**, 4 (2017)
8. I.Pane, W.Hansen, Mater. Struct., **35**, 2 (2002)
9. S.Altoubat, M.T.Junaid, M.Leblouba, D.Badran, Cem. Concr. Compos., **79**(2017)
10. J.Xin, G.Zhang, Y.Liu, Z.Wang, Wu Z., Constr. Build. Mater., **192** (2018)
11. Y.Ding, C.Zhang, M. Cao, Construction and Building Materials, **113**,15(2016)
12. B.Chen, J.Liu, Cement and Concrete Research, **34**, 6(2004)
13. P. Kalifa, G. Chéné, C. Gallé, Cement & Concrete Research, **31**, 10(2001)
14. Y.Ding, C.Azevedo, J. Aguiar, S. Jalali, Construction & Building Materials, **26**, 1(2012)

15. The National Standards Compilation Group of the People's Republic of China, GB 50010-2010. "Common Portland cement." Beijing: China Standards Press (2015)

High-Performance Control Technology of Buck Inverter Used for Super-Precision Machining of Composite Materials

*En-Chih Chang**, *Hung-Liang Cheng, and Chun-An Cheng*

Department of Electrical Engineering, I-Shou University, Kaohsiung City, Taiwan, R.O.C.

Abstract. An effective technique, genetic-feedforward sliding mode-fractional PI controlled buck inverter used for super-precision machining of composite materials is proposed. The obstruction using sliding mode control (SMC) is due to the strong chattering that severely limits its practical applicability. The chattering yields high voltage distortion in buck inverter output, thus degrading stability and reliability of super-precision machining of composite materials. The modification structure to fractional PI has been established through the plant extending way so that the chattering is diminished and better flexibility in adjusting system response can be provided. The feedforward compensator improves the dynamics response further. The genetic algorithm (GA) can be adopted for determining optimal fractional proportional-integral (FPI) parametric values. With this control technique, the TI microprocessor-based buck inverter is implemented, and then experiments illustrate that the presented technique produces less steady-state inaccuracy, chattering attenuation, loading interference rejection and parametric variation removal.

1 Introduction

Super-precision machining of composite materials (such as aluminum matrix composite, Cu-based composite, and magnesium matrix composite materials) needs stable and high-quality power sources [1, 2], i.e., 1. inverter wave shape with small amounts of total harmonic distortion even under rectifier loading. 2. rapid transience against step changes in load. 3. zero steady-state errors. High performance buck inverter can provide these requirements by employing closed-loop control techniques [3, 4]. Early works on inverter control can be found in literature, for example, PI control [5], multiple feedback strategy [6], dead-beat method [7], and Fourier approach [8]. Although above schemes have presented the satisfactory results, both the low total harmonic distortion of output voltage and fast transient response cannot simultaneously be achieved. The SMC is capable of making a control system robust against internal parameter variations and external load disturbances [9, 10]. A number of SMC method developed for inverters has been done [11, 12]. However, it is a pity that the chattering always exists. The chattering may damage inverter parts and cause highly distorted wave shape even if a SMC had incorporated a feedforward path so as to improve the dynamics response [13]. This study develops a plant extending architecture for designing feedforward sliding mode-fractional PI (FSM-FPI) controller with reduced chattering and higher flexibility. A FSM output becomes the plant input that is unable to endure a chatter. In order to overcome the chattering, a fractional-order extended plant, which implies the FPI is introduced [14-16]. Henceforth, the chattering happens at the input of the extended plant. Such chattering locates at system lower powered zone and therefore is inoffensive. It's important to note that optimal FPI parametric values maybe not easy for selection in the presence of various perturbations [17-26]. A GA conception firstly developed in 1975 by John Holland of the University of Michigan [27]. Then, De. Jong expands its application in function optimum [28-32]. The FPI parametric optimum is capable of obtaining determination via GA. Using proposed method, a buck inverter is very suitable for the application of super-precision machining of composite materials, yielding many advantages, such as low total harmonic distortion wave shape, satisfactory voltage regulation, reduced chattering, reduced steady-state inaccuracy and high-accuracy of the super-precision machining.

2 Problem statement

Consider the design of a SMC for controlled system below.

$$\dot{\tilde{x}} = A\tilde{x} + bu + f \tag{1}$$

where $\tilde{x}$ symbolizes vector of variable errors, A denotes state transition matrix, b represents input matrix, u stands for control input to the system, and f signifies the disturbance.

The $\tilde{x}$ is given by

$$\tilde{x} = x_d - x \tag{2}$$

* Corresponding author: enchihchang@isu.edu.tw

where x_d and x is the vector of references and state vector, respectively.

The sliding surface governs the dynamic system that can be given by

$$\sigma = c\tilde{x} \tag{3}$$

where c is constant.

A SMC control law is

$$u = u_f + u_{eq} + u_s \tag{4}$$

where u_f =feedforward compensation control, u_{eq} =equivalent control, and u_s =switching control.

The dynamics in sliding action is defined as

$$\dot{\sigma} = c\dot{\tilde{x}} = 0 \tag{5}$$

From (1) and (5), the equivalent control can be obtained as

$$u_{eq} = -(cb)^{-1}cA\tilde{x} \tag{6}$$

By the use of the $\sigma\dot{\sigma} < 0$, the switching control can be derived as

$$u_s = -(cb)^{-1}\rho sgn(c\tilde{x}) \tag{7}$$

where $\rho > 0$.

Evidently, the (7) uses an sgn function, the chattering will occur in practice. One solution reduces the chattering by means of section 3 design.

3 Control technology design

Suppose a control-system dynamics be

$$\dot{x} = Ax + bu + f \tag{8}$$

where x signifies a vector of system state, A stands for a state transition matrix, b represents an input matrix, u denotes a control input, and f is the disturbance.

To generate the expanded system states, a new vector can be constructed as

$$\dot{x} = \Psi \tag{9}$$

where Ψ is a vector.

Substituting (9) into (8) yields

$$\Psi = Ax + bu + f \tag{10}$$

The u is designed in the following.

$$u = ((K_p s^{\lambda} + K_i)\frac{1}{s^{\lambda}})\tilde{x} \tag{11}$$

where k_p and k_i stand for two gains.

Hence, there is a relationship as follows:

$$\frac{u}{u_a} = \frac{1}{s^{\lambda}} \tag{12}$$

where a FPI architecture can be written as

$$u_a = (K_p s^{\lambda} + K_i)\tilde{x} \tag{13}$$

From (2) and (9)-(13), the error equations are formulated by

$$\begin{aligned} \dot{\tilde{x}} &= \dot{x}_d - \Psi \\ \ddot{\tilde{x}} &= A^2\tilde{x} - Abu - bu_a - A^2 x_d + \ddot{x}_d - Af - \dot{f} \end{aligned} \tag{14}$$

The sliding function of the extended system can be defined as

$$\sigma = c\tilde{x} + \dot{\tilde{x}} \tag{15}$$

where c is constant

The following inequality can guarantee the existence of sliding mode

$$\lim_{\sigma \to 0} \sigma\dot{\sigma} < 0 \tag{16}$$

The conditions with K_p and K_i for satisfying (16) yield, and obviously the two parameters K_p and K_i are decided in a wide range. The optimum of the two parameters is sought blindly, and the local optimum may occur. In order to search the optimal parameters, a global optimum tool, namely, GA is employed for FSM-FPI controller.

The genetic algorithm procedure for FSM-FPI controller is outlined in the following.

1. Encoding two parameters K_p and K_i.
2. Producing an initial generation.
3. Computing the fitness function.
4. Remaking current population through a selected mechanism.
5. Crossover is performed by a couple of members among new generation.
6. Using the mutation probability, the mutation is done in every number among new generation.
7. The maximum value of a fitness function is reserved in the old population to the current population.
8. Iteratively execute above steps until the acceptable solution appears.

4 Design example

For clarifying design technology in the section 3, a structure of a buck inverter is presented in Fig. 1 and its equivalence circuitry is plotted in Fig. 2. A load, which allows linearity or nonlinearity is tested so that buck inverter performance can be validated. A buck inverter output is design to track sine-reference v_{ref}.

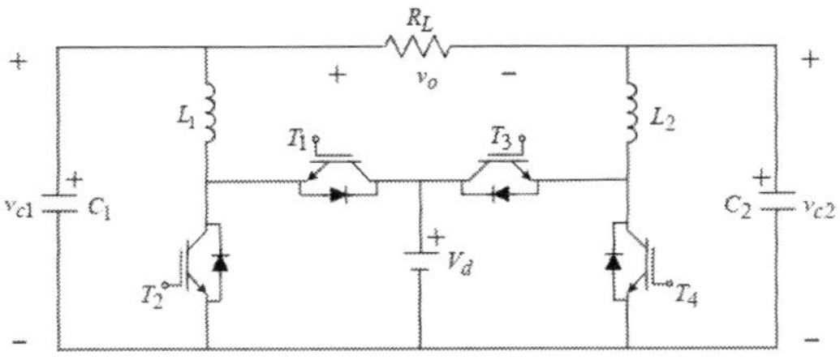

Fig. 1. Circuit diagram of buck inverter

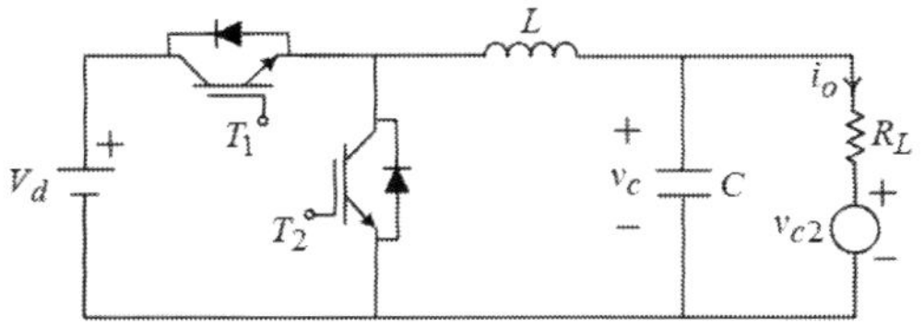

Fig. 2. Buck inverter's equivalence circuitry

https://doi.org/10.1051/matecconf/201926403001

From the Fig. 1, the state space equation can be expressed as

$$\begin{bmatrix} \dot{v}_c \\ \ddot{v}_c \end{bmatrix} = \begin{bmatrix} 1 & 0 \\ -\dfrac{1}{LC} & -\dfrac{1}{R_L C} \end{bmatrix} \begin{bmatrix} v_c \\ \dot{v}_c \end{bmatrix} + \begin{bmatrix} 0 \\ \dfrac{V_d}{LC} \end{bmatrix} u + \begin{bmatrix} 0 \\ \dfrac{\dot{v}_{c2}}{R_L C} \end{bmatrix} \quad (17)$$

where $v_c = v_{c1}$, $L_1 = L$ and $C_1 = C$.

Then, the error differential equations expressing the system yield

$$\tilde{x}_1 = v_{ref} - v_c \quad (18)$$

$$\dot{\tilde{x}}_1 = \dot{v}_{ref} - \dot{v}_c = \tilde{x}_2 \quad (19)$$

$$\begin{aligned} \dot{\tilde{x}}_2 = \tilde{x}_3 &= \ddot{v}_{ref} - \ddot{v}_c \\ &= \ddot{v}_{ref} - a_1\tilde{x}_1 - a_2\tilde{x}_2 - bu + N + \Gamma \end{aligned} \quad (20)$$

$$\begin{aligned} \dot{\tilde{x}}_3 &= \dddot{v}_{ref} - \dddot{v}_c \\ &= \dddot{v}_{ref} - a_1\tilde{x}_2 - a_2\tilde{x}_3 - bK_p(s^{\lambda}\tilde{x}_1) - bK_i\tilde{x}_1 + \dot{N} + \dot{\Gamma} \end{aligned} \quad (21)$$

where $a_1 = \dfrac{1}{LC}$, $a_2 = \dfrac{1}{R_L C}$, $b = \dfrac{V_d}{LC}$, $N = \dfrac{\dot{v}_{c2}}{R_L C}$, and $\Gamma = a_1 v_{ref} + a_2\dot{v}_{ref} + \ddot{v}_{ref}$.

The sliding surface is chosen as

$$\sigma = m_1\tilde{x}_1 + m_2\tilde{x}_2 + \tilde{x}_3 \quad (22)$$

where m_1 and m_2 are constants.

Then,

$$\begin{aligned} \dot{\sigma} = {} & m_1\tilde{x}_2 - m_2 a_1\tilde{x}_1 - m_2 a_2\tilde{x}_2 - bu + +N + \Gamma \\ & - a_1\tilde{x}_2 - a_2\tilde{x}_3 - bK_p(s^{\lambda}\tilde{x}_1) - bK_i\tilde{x}_1 + +\dot{N} + \dot{\Gamma} \end{aligned} \quad (23)$$

where $a_1 = a_{10} + \Delta a_1$, $a_2 = a_{20} + \Delta a_2$, and $b = b_0 + \Delta b$.

Redefine the control function u_{ex} as

$$u_{ex} = u_{ex\_f} + u_{ex\_eq} + u_{ex\_s} \quad (24)$$

The equivalent control u_{ex_eq} is derived by underneath operation:

$$\dot{\sigma}\big|_{a_1=a_{10},\, a_2=a_{20},\, b=b_0,\, \Gamma=0} = 0 \quad (25)$$

The feedforward compensator u_{ex_f} is employed to enhance the tracking performance as follows.

$$\dot{\sigma}\big|_{a_1=a_{10},\, a_2=a_{20},\, b=b_0,\, N=0} = 0 \quad (26)$$

The sliding control u_{ex_s} is used to suppress the parameter variations and load disturbances that can be chosen as

$$u_{ex\_s} = \Phi_1\tilde{x}_3 + \Phi_2 \quad (27)$$

An inequality depicts sliding existence as follows:

$$\sigma\dot{\sigma} < 0 \quad (28)$$

Substituting (25)-(27) into (28), the parameters K_p, K_i, Φ_1, and Φ_2 will be obtained to satisfy the (28). To find the optimum parameters K_p and K_i, the GA can be adopted for searching global optimization parameters. The process of optimizing the two parameters describes concretely as follows: (i) Two parameters should be encoded into a string of eight bits. (ii) An initial generation $P(0) = 150$ is produced randomly. (iii) High crossover probability has fast search, and thus $P_c = 0.8$. (iv) Generally, the probability of mutation is low, and therefore $P_m = 0.045$. (v) Reserving maximum value of a fitness function in the old population to the current population. (vi) Once the seek time is enough long, the solution is obtained.

5 Experimental results

The robustness to buck inverter will be investigated via the experiments. The parameters of the buck inverter used for the proposed method display as Table 1.

Table 1. System parameters

DC voltage	200 V
Filter capacitor	2 μF
Filter inductor	0.5 mH
Switching frequency	15 kHz
Output voltage and frequency	110 V_{rms}, 60 Hz

Figure 3 shows the proposed method under a resistive load 12 Ω condition. For the purpose of testing transience further, the step-changing load circumstances while the resistive load is abruptly connected, as shown in Fig. 4. We can find that the developed buck inverter has slight voltage droop and then rapid retrieve in face of step-changing load. A nonlinear load is composed of a full-wave rectified with a parallel resistor, 35 Ω and capacitor filter, 270 μF. Although load current shown in Fig. 5 is distorted highly, the output-voltage wave shape for the presented technique exhibits a low voltage %THD (total harmonic distortion) of 1.16%. Fig. 6 illustrates that the developed buck inverter can allow random variations of LC filter parameters, from 15% to 250% of nominal values in the presence of resistive load 12 Ω. The phase plane of the proposed method under nonlinear load is shown in Fig. 7. It is very clear that the proposed technique effectively not only alleviates the chatter but also improves transient response due to the feedforward compensator.

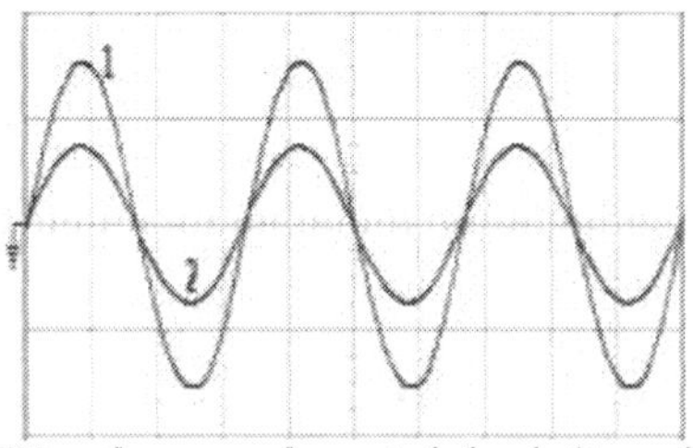

Fig. 3. Waveforms under rated load (1: voltage; 2: current) (vert.: 100 Volts/division and 20Ampere/division, hor.: 5msec/division)

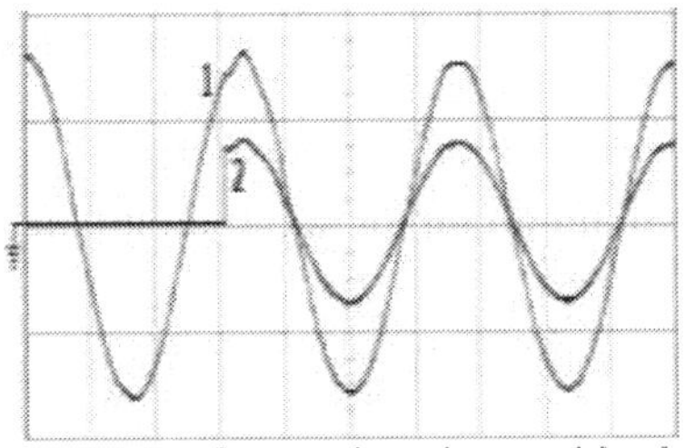

Fig. 4. Output waveforms when the rated load abruptly applied (1: voltage; 2: current) (vert.: 100 Volts/division and 20Ampere/division, hor. : 5msec/division)

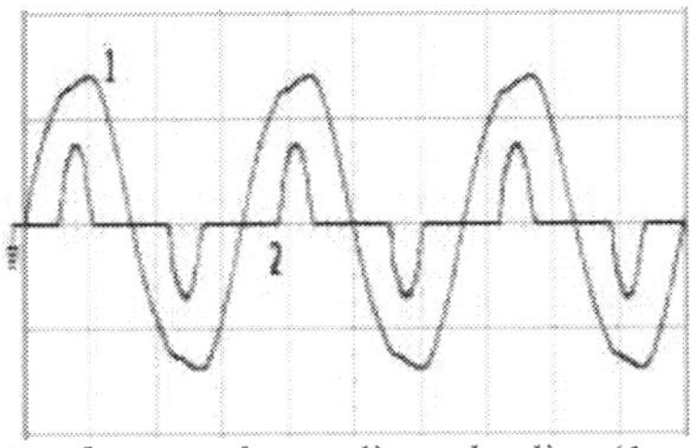

Fig. 5. Waveforms under nonlinear loading (1: voltage; 2: current) (vert.: 100 Volts/division and 20Ampere/division, hor. : 5msec/division)

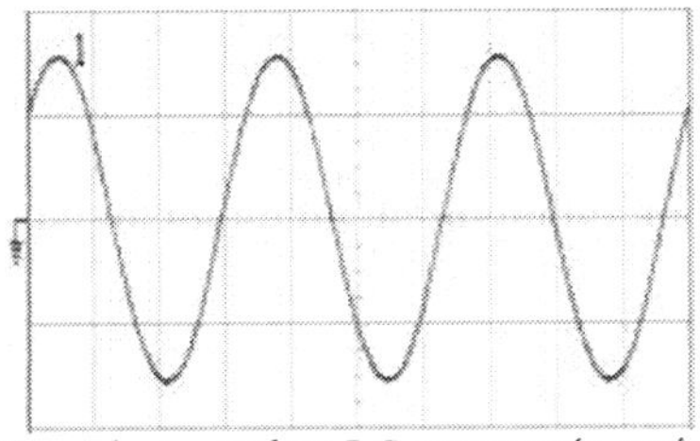

Fig. 6. Wave shape against LC parametric variations (1: voltage) (vert.: 100 Volts/division, hor.: 5msec/division)

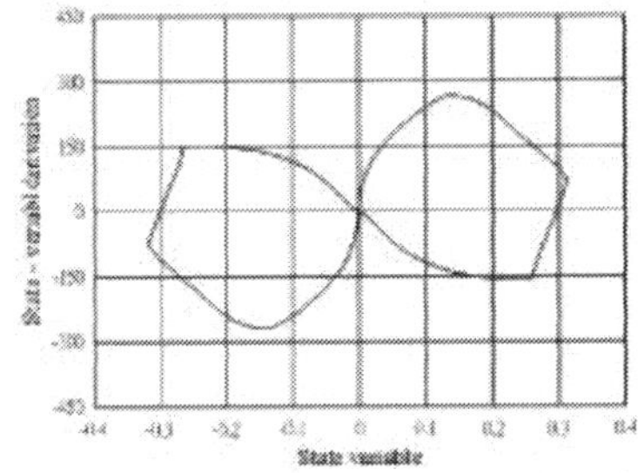

Fig. 7. Phase plane for the proposed method

6 Conclusions

This study has proposed genetic-feedforward sliding mode-fractional PI controller suitable for buck inverter used in super-precision machining of composite materials, to improve the dynamic response, to alleviate the chattering, to optimize the FPI parameters and to enhance machining stability. FSM not only has the merits of SMC but also improves the dynamics response by feedforward compensator. The extended plant method alleviate the chattering due to the FPI combined FSM. Furthermore, the GA is employed in FSM-FPI controller to optimize the FPI parameters. Thus, the buck inverter yields the good voltage regulation, fast dynamic response, and low THD output voltage. According to the experimental results and theoretic analysis, the proposed buck inverter is successfully demonstrated, thus supplying the reliability of super-precision machining of composite materials.

Acknowledgment

The authors would like to thank the Ministry of Science and Technology, Taiwan, R.O.C. for financially supporting this research under Contract No. MOST 107-2221-E-214-006.

References

1. A. Ochsner, and H. Altenbach, Machining, Joining and Modifications of Advanced Materials, Springer, 2016.
2. J. P. Davim, Machining of Metal Matrix Composites, Springer, 2012.
3. R. Yuriy, R. Sergey, C. Evgeny, and V. Pavel, Power electronics basics: operating principles, design, formulas, and applications, CRC Press, Boca Raton, FL, USA, 2016.
4. H. Ma, and S. L. Han, "Buck inverter with sliding mode control strategy", in *Proceedings of the 2004 IEEE Industrial Electronics Conference*, 2004, pp.592-597.
5. H. I. Cha, S. S. Kim, M. G. Kang, and Y. H. Chung, "Real-time digital control of PWM inverter with PI compensator UPS," in *Conf. Rec. IEEE-IAS Annu. Meeting*, 1990, pp. 1124-1128.
6. S. L. Jung, H. S. Huang, M. Y. Chang, and Y. Y. Tzou, "DSP-based multiple-loop control strategy for single-phase inverters used in ac power sources," *IEEE PESC'97*, vol. 1, 1997, pp. 706-712.
7. A. Kawamura, T. Haneyoshi, and R. G. Hoft, "Dead beat controlled PWM inverter with parameter estimation using only voltage sensor," in *Proc. IEEE PESC*'86, 1986, pp. 576-583.
8. A. V. Jouanne, P. N. Enjeti, and D. J. Lucas, "DSP control of high-power UPS system feeding nonlinear loads," *IEEE Trans. Ind. Electron*., vol. 43, pp.121-125, 1996.
9. U. Itkis, Control systems of variable structure. New York: Wiley, 1976.
10. K. D. Young, V. I. Utkin, and U. Ozguner, "A control engineer's guide to sliding mode control," *IEEE Trans. Contr. Syst. Technol*., vol. 7, pp. 328-342, 1999.
11. S. Vaidyanathan, and C. H. Lien, Applications of Sliding Mode Control in Science and Engineering, Springer, New York, 2017.
12. G. Bartolini, L. Fridman, A. Pisano, and E. Usai, Modern Sliding Mode Control Theory, Springer-Verlag, Berlin, 2008.
13. S. L. Jung, and Y. Y. Tzou, "Discrete feedforward sliding mode control of a PWM inverter for

sinusoidal output waveform synthesis," in *Proc. IEEE PESC'94*, 1994, pp.552-559.
14. I. Podlubny, "Fractional-order Systems and $PI^{\lambda}D^{\mu}$–Controllers," *IEEE Transactions on Automatic Control*, vol. 44, no. 1, pp. 208-214, 1999.
15. C. N. Zhao, and X. D. Zhang, "The Application of Fractional Order PID Controller to Position Servomechanism," *Proc. World Congress on Intelligent Control and Automation*, 2008, pp. 3380-3383.
16. H. M. Fan, Y. Sun, and X. B. Zhang, "Research on Fractional Order Controller in Servo Press Control System," *Proc. Int. Conf. Mechatronics and Automation*, 2007, pp. 2934- 2938.
17. I. Podlubny, L. Dorcak, and I. Kostial, "On Fractional Derivatives, Fractional-order Dynamic Systems and $PI^{\lambda}D^{\mu}$-Controllers," *Proc. Int. Conf. Decision and Control*, Vol. 5, 1997, pp. 4985-4990.
18. I. Podlubny, "Geometric and Physical Interpretation of Fractional Integration and Fractional Differentiation," *Fractional Calculus and Applied Analysis*, Vol. 5, No. 4, 2002, pp. 367-386.
19. Y. Q. Chen; I. Petras, and D. Xue, "Fractional Order Control- A Tutorial," *Proc. Int. Conf. American Control Conference*, 2009, pp. 1397-1411.
20. Q. S. Zeng, G. Y. Cao, and X. J. Zhu, "The Effect of the Fractional-order Controller's Orders Variation on the Fractional-order Control Systems," *Proc. Int. Conf. Machine Learning and Cybernetics*, 2002, pp. 367-372.
21. C. Ma, and Y. Hori, "Fractional Order Control and Its Application of $PI^{\alpha}D$ Controller for Robust Two-Inertia Speed Control," *Proc. Int. Conf. Power Electronics and Motion Control*, Vol. 3, 2004, pp. 1477-1482.
22. C. B. Ma, and Y. Hori, "Backlash Vibration Suppression in Torsional System Based on the Fractional Order Q-Filter of Disturbance Observer," *Proc. IEEE Int. Workshop Advanced Motion Control*, 2004, pp. 577-582.
23. N. M. F. Ferreira, and J. A. T. Machado, "Fractional-Order Hybrid Control of Robotic Manipulators," *Proc. Int. Conf. Advanced Robotics*, 2003, pp. 393-398.
24. I. Petras, and B. M. Vinagre, "Practical Application of Digital Fractional-Order Controller to Temperature Control," Acta Montanistica Slovaca, Vol. 7, No. 2, 2002, pp. 131-137.
25. F. C. Clark, "PID algorithms and their computer implementation," *IEEE Trans. Inst. Meas. Contr.*, vol. 6, pp. 305-316, 1984.
26. W. K. Ho, K. W. Lim, and W. Xu, "Optimal gain and phase margin tuning for PID controllers," *Automatica*, vol. 34, pp. 1009-1014, 1998.
27. J. H. Holland, *Adaptation in Natural and Artificial Systems*. Ann Arbor, Michigan: The University of Michigan Press, 1975.
28. T. R. Smith, and K. A. De Jong, "Genetic algorithms applied to the calibration of an information driven model of us migration patterns," in *Proc. Conf. Modeling and Simulation*, 1981, vol. 12, pp. 955-959.
29. K. A. De Jong, W. M. Spears, and D. F. Gordon, "Using genetic algorithm for concept learning," in *Proc. Conf. Machine Learning*, 1993, vol. 13, no. 2, pp. 161-188.
30. M. A. Ahmed, and K. A. De Jong, "Function approximator design using genetic algorithms," in *Proc. Conf. IEEE Evol. Comput.*, 1997, pp. 519-524.
31. S. N. Sivanandam, and S. N. Deepa, Introduction to Genetic Algorithms, Springer, 2008.
32. Kramer, Genetic Algorithm Essentials, Springer, 2017.

Accuracy of Real Space Cluster Expansion for Total Energies of Pd-rich PdX (X=Rh, Ru) Alloys, based on Full-Potential KKR Calculations for Perfect and Impurity Systems

Mitsuhiro Asato[1,2,a], Chang Liu[3], Nobuhisa Fujima[4], Toshiharu Hoshino[5], Ying Chen[6], and Tetsuo Mohri[7]

[1]*National Institute of Technology, Niihama College, Niihama, Ehime, 792-8580, Japan*
[2]*Department of Chemistry and Biochemistry, University of California, Los Angeles, Los Angeles, California 90095, United States*
[3]*National Institute for Materials Science, Tsukuba, Ibaraki, 305-0047, Japan*
[4]*Graduate School of Engineering, Shizuoka University, Hamamatsu, 432-8561, Japan*
[5]*Faculty of Engineering, Shizuoka University, Hamamatsu, 432-8561, Japan*
[6]*Graduate School of Engineering, Tohoku University, Sendai, 980-8579, Japan*
[7]*Institute for Materials Research, Tohoku University, Sendai, 980-8577, Japan*

Abstract. We study the accuracy and convergence of the real space cluster expansion (RSCE) for the total energies of the Pd-rich PdX (X=Ru, Rh) alloys, which are used to study the phase stability and phase equilibria of the Pd-rich PdX alloys. In the present RSCE, the X atoms of minor element are treated as impurities in Pd. The *n*-body interaction energies (IEs) among X impurities in Pd, being used in the expansion of the total energies of the Pd-rich PdX alloys, are determined uniquely and successively from the low body to high body, by the full-potential Korringa-Kohn-Rostoker (FPKKR) Green's function method (FPKKR) for the perfect and impurity systems (Pd-host and X_n in Pd, n=1~4), combined with the generalized gradient approximation in the density functional theory. In the previous paper, we showed that the RSCE, in which the perturbed potentials due to the insertion of X_n impurities in Pd were redetermined self-consistently up to the first-nearest neighboring (*nn*) host atoms around X_n impurities, reproduce fairly well (the error of ~ 0.2mRy per atom) the FPKKR-band-calculation result of the ordered Pd_3Rh alloy in $L1_2$ structure, but a little wrongly (the error of ~ 0.7mRy per atom) for the ordered Pd_3Ru alloy in $L1_2$ structure. In the present paper, we show that this small RSCE error for the Pd_3Ru alloy is corrected very well (from ~ 0.7mRy to ~ 0.1mRy per atom) by enlarging the self-consistent region for the perturbed potentials up to the 2nd-*nn* host atoms around Ru_n impurities in Pd. We also clarify the correction for each value of the *n*-body (n=1~ 4) IEs.

1 Introduction

The knowledge of interaction energies of point defects in solids is indispensable for the understanding of many basic physical processes, such as diffusion, short-range order, and phase diagrams, *etc*. We have shown that the type of phase diagrams of A-rich AX binary alloys of host (A) and impurity (X) elements, such as segregation, solid solution, and order known experimentally, may be very well discriminated by use of the sign and magnitude of the first-nearest neighboring (*nn*) 2-body (X_2) interaction energies (IEs) in A [1]. We have also succeeded in describing the relative stabilities among different atomic structures by using the 2-body and 3-body IEs among X impurities in A; the stabilities of the ordered Al-rich AlX (X=Sc~V) alloys, the stability of AlMn quasicrystal, and the shapes of agglomerates of X impurities in the initial stage of the formation of the so-called Guinier-Preston zones of Al-rich $Al_{1-c}X_c$ (X=Cu, Zn; c<0.05) alloys, such as the (001) disc in Al-rich AlCu alloys and the spherical shape in Al-rich AlZn alloys [2-5]. All of these stabilities were obtained by the present real space cluster expansion (RSCE) [2-7]. The *n*-body (n=1~4) IEs among X impurities in A were determined by the *ab-initio* calculations based on the full-potential Korringa-Kohn-Rostoker Green's function method (FPKKR) for the perfect and impurity systems (A-host and X_n impurities in A, n=1~4), combined with the generalized gradient approximation in the density functional theory [8, 9].

We recently succeeded in quantitatively reproducing the X-concentration dependence of the observed solvus temperatures of the Pd-rich PdX (X=Ru, Rh) alloys in fcc structure [10], by the free energy calculations based on the cluster variation method (CVM) [11-13]. The internal energies in the free energies of the Pd-rich PdX (X=Ru, Rh) alloys, were calculated by use of the present RSCE from a dilute limit [2-7]. We have already studied the distance dependence of the 2-body IEs, the local lattice distortion for the 2-body IEs, existing even at 0K, and both the temperature effects for (1) the thermal vibration by the Debye-Grüneisen model and (2) the electron excitation due to the Fermi-Dirac (FD) distribution [11-13]. We very recently showed that the present RSCE

[a] Corresponding author: asato@sci.niihama-nct.ac.jp

reproduces fairly well the FPKKR-band-calculation results for total energies of the ordered Pd_3X (X=Ru, Rh) alloys in $L1_2$ structure (X-concentration = 25%) by including the 2-body IEs up to the 20th-*nn* pair (X_2) and the 3-body IEs up to the clusters (X_3) in two cubic cells in fcc structure [14]. We obtained the nice agreement (the error of ~ 0.2mRy per atom) for the Pd_3Rh alloy, while slightly wrong result (the error of ~ 0.7mRy per atom) for the Pd_3Ru alloy. We considered that this difference between the RSCE results for X=Ru and for X=Rh might be due to the difference of the delocalization of the perturbed potentials in the X_n impurity systems, depending on the delocalization of 4d orbitals of X impurities. It is noted that in these calculations, the perturbed potentials due to the insertion of X impurities were self-consistently redetermined up to the first-*nn* host atoms around X_n impurities in Pd, as in the usual KKR calculations for the impurity systems [9].

The purpose of the present paper is to study how the RSCE results are improved by enlarging the self-consistent (sc) region for the perturbed potentials around impurities in Pd. In Section 2, we describe the calculation procedures for the *ab-initio* calculations based on the full-potential Korringa-Kohn-Rostoker (FPKKR) method with the electron excitation effect due to the FD distribution and the present RSCE from a dilute limit. In Section 3, we show that, by enlarging the sc-region for the perturbed potentials up to the 2nd-*nn* host atoms around X impurities in Pd, the discrepancy between the RSCE and FPKKR-band-calculation results for the total energy of the ordered Pd_3X alloy in $L1_2$ structure is corrected very well (from ~ 0.7mRy to ~ -0.1mRy per atom) for X=Ru, while slightly (from ~ 0.2mRy to ~ 0.1mRy per atom) for X=Rh. As results, the absolute error of the RSCE becomes as small as ~ 0.1mRy per atom for both the ordered Pd_3X (X=Ru, Rh) alloys in $L1_2$ structure. We also clarify the correction for each value of the *n*-body IEs (n=1~4). Section 4 summarizes the main results of the present paper and discusses the future problems.

2 Calculation Method

2.1 Total Energies of Impurity Systems

The calculations for the total energy of X (=Ru, Ru) impurities in Pd are based on the density functional theory (DFT) in the generalized gradient approximation (GGA) [8, 9]. In order to solve the Kohn-Sham equations, we use a multiple scattering theory in the form of the KKR Green's function method for the full-potentials (FP). We use the screened version of FPKKR band calculations for Pd host, which simplifies very much the numerical calculations by introducing the short-range structural Green's functions [15, 16]. In order to simplify the total energy calculations in the GGA formalism, we use the electronic densities obtained self-consistently by the local spin density approximation in the DFT. The accuracy of the present GGA calculations are discussed in Refs.8 and 9. It was already shown in Ref. 17 that the introduction of the FD distribution into *ab-initio* Green's function calculations is very useful to simplify the total energy calculations. Based on a contour integration in the complex energy plane we evaluate the residues at a few Matsubara frequencies, while the remaining integral is analyzed by the Sommerfeld expansion. The lattice parameter of Pd host and T_{FD} for the electron excitation due to the FD distribution are fixed at 7.40a.u. and 800K, respectively, in all the present calculations.

The advantage of the Green's function method is that the embedding of point defects in an otherwise ideal crystal is described correctly by introducing the host Green's function, differently from the usual supercell and cluster calculations. In order to calculate the total energies of the impurity systems, the perturbed potentials due to the insertion of impurities are usually redetermined up to the 1st-*nn* host atoms around impurities [9]. However, in the present study, we redetermine the perturbed potentials up to the 2nd-*nn* host atoms around impurities, in order to examine the delocalization of the perturbed potentials.

2.2 Real Space Cluster Expansion for Total Energies of Pd-rich PdX (X=Ru, Rh) Alloys

A characteristic feature of the present RSCE is that all the terms in the present RSCE are determined uniquely and successively from the low body to high body, by the combination of total energies in different atomic arrangements of X atoms in Pd. It is noted that the atoms of minor X element are considered as impurities in Pd. The RSCE of the total energy (per unit cell (4 atoms)) of Pd_3X alloy in $L1_2$ structure (Fig.1 (a)) is written as follows:

$$E^{\text{band}}_{\text{Pd}_3\text{X}} = 4E^{\text{band}}_{\text{Pd}} + \Delta E^{\text{1-body}} + \Delta E^{\text{2-body}} + \Delta E^{\text{3-body}} + \Delta E^{\text{4-body}} + \Delta E^{\text{many-body IEs beyond 4-body}} \quad (1)$$

where the first term is 4 × total energy of Pd atom in Pd (0-body), the second term the total energy change due to the insertion of a single X impurity in Pd (1-body), the 3rd~5th terms correspond to the total contributions from the *n*-body (n=2~4) IEs among X_n impurities in Pd (2~4-body), respectively, and the last term is the sum total of many-body IEs beyond the 4-body. Since the first (0-body) and second (1-body) energies are the quantities characteristic of Pd-host and a single X impurity in Pd, respectively, these components do not depend on the atomic arrangement of X impurities in Pd. It is noted that the atomic arrangement for X impurities in Pd, at low temperatures, are determined by the minimization of the total sum of the *n*-body IEs ($n \geq 2$). The definition and the calculational procedure for each term in eq. (1) are described in Refs. 4 and 14.

3 Accuracy of Real Space Cluster Expansion for Total Energies of Pd_3X (X=Ru, Rh) Alloys in $L1_2$ structure and Delocalization of Perturbed Potentials due to X Impurities

https://doi.org/10.1051/matecconf/201926403002

In the previous study we found that the present RSCE reproduces fairly well (the error of ~ 0.2mRy per atom) the band calculation result for the ordered Pd_3Rh alloy in $L1_2$ structure, while a little wrongly (the error of ~ 0.7mRy per atom) for the Pd_3Ru alloy in $L1_2$ structure [14]. In the previous study, however, we redetermined self-consistently the perturbed potentials due to the insertion of X impurities in Pd, up to the 1st-*nn* host atoms around X impurities in Pd. Thus, it may be easily understood that the RSCE error becomes large for the impurity systems in which the perturbed potentials are delocalized beyond the 1st-*nn* host atoms around X impurities in Pd; the perturbed potentials due to the insertion of X impurities in Pd may be delocalized for X=Ru than for X=Rh. It is noted that the nuclear charge of Ru-element is smaller than that of Rh-element so that the 4d-orbitals of X impurities in Pd is delocalized for X=Ru than for X=Rh. In the present study, we examine the delocalization of the perturbed potentials by enlarging the sc-region for the perturbed potentials up to the 2nd-*nn* host atoms around impurities in Pd.

For the total energies of the ordered Pd_3X (X=Ru, Rh) alloys in $L1_2$ structure (Fig. 1(a)), we include the 2-body IEs up to the 20th-*nn*, the 3-body IEs for the X_3 clusters shown in Fig. 1(b), and the 4-body IEs for the X_4 clusters shown in Fig.1(c), as discussed in Ref. 14. Table 1 shows the calculated results by the present RSCE (eq. (1)) from a dilute limit: the total energy of Pd-host, the total energy changes due to the insertion of a single X impurity in Pd, and the total contributions from the *n*-body (*n*=2~4) IEs. For comparison, we also show (in the parentheses) the calculated results with the sc-region for the perturbed potentials up to the 1st-*nn* host atoms around impurities in Pd, being already shown in Ref. 14.

We first find, in the results by enlarging the sc-region for the perturbed potentials, the discrepancy between the RSCE and the FPKKR-band-calculation results for the Pd_3X is satisfactorily corrected (from 0.71mRy to -0.06 mRy per atom) for X=Ru, while slightly (from 0.17mRy to 0.05mRy per atom) for X=Rh because the band-calculation result for the Pd_3Rh is reproduced very well by the RSCE even with the sc-region for the perturbation potentials up to the 1st-*nn* host atoms around impurities in Pd, as shown in the parenthesis.

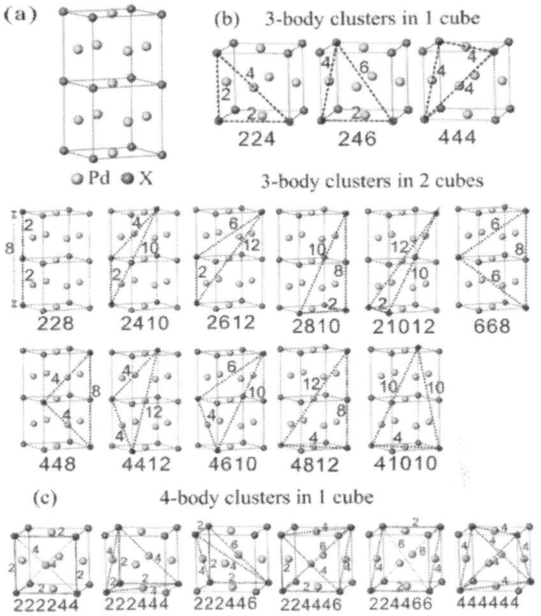

Figure 1. Atomic arrangement for X impurity clusters of the 3-body and 4-body IEs, used in the RSCE of the total energies of Pd_3X (X=Ru, Rh): (a) $L1_2$ structure (4 atoms in unit cell), (b) X_3 clusters for 3-body IEs, and (c) X_4 clusters for 4-body IEs. For 3-body IEs, there are the three (eleven) kinds of X_3 clusters in one cubic cell (two cubic cells) in fcc structure. For 4-body IEs, there are six kinds of X_4 clusters in one cubic cell in fcc structure. For example, the 224-cluster (the smallest X_3-cluster for 3-body IEs) means that this cluster consists of two 2nd-*nn* and one 4th-*nn* pairs.

Table 1. Convergence of the RSCE for the total energies (in Ry) of (a) Pd_3Ru and (b) Pd_3Rh in $L1_2$ structure. The perturbed potentials due to the insertion of impurities are redetermined up to the 2nd-*nn* host atoms around impurities in Pd. The values in the parentheses are the calculated results in which the perturbed potentials due to the insertion of impurities are redetermined up to the 1st-*nn* host atoms around impurities in Pd.

	(a) Toal energies of Pd_3Ru (Ry)	(b) Total energies of Pd_3Rh (Ry)
0-body $[4 \times E^{Pd}_{band}]$	-40374.470033 [4×−10093.617508]	-40374.470033 [4×−10093.617508]
up to 1-body $[\Delta E^{1\text{-body}}]$	-39344.445216 (-39344.444854) [1030.024816 (1030.025179)]	-39851.104786 (-39851.104688) [523.365247 (523.365344)]
up to 2-body $[\Delta E^{2\text{-body}}_{int}]$	-39344.425543 (-39344.423629) [0.019674 (0.021225)]	-39851.103386 (-39851.102828) [0.001400 (0.001861)]
up to 3-body $[\Delta E^{3\text{-body}}_{int}]$	-39344.432149 (-39344.4289813) [-0.006607 (-0.005352)]	-39851.104065 (-39851.103456) [-0.000679 (-0.000628)]
up to 4-body $[\Delta E^{4\text{-body}}_{int}]$	-39344.431167 (-39344.428093) [0.000983 (0.000888)]	-39851.103919 (-39851.103441) [0.000147 (0.000015)]
exact error	-39344.430939 [-0.000228 (0.002845)]	-39851.104114 [0.000195 (0.000673)]
error/atom	-0.000057 (0.000711)	0.000049 (0.000168)

The corrections for the total contributions from the n-body (n=2~4) IEs are also very small (-0.46, -0.05, and 0.13mRy per cell) for the Pd_3Rh alloy, while somehow large (-1.55, -1.26, and 0.10mRy per cell) for the Pd_3Ru alloy. We also find that the correction for the 1-body IE is as small as -0.36(-0.10)mRy for X=Ru (Rh). It is noted that the absolute value of the 1-body IE is very large because it is the total energy change due to the replacement of one Pd-host atom with one X impurity, while the n-body IEs (n=2~4) are small because the n-body (n=2~4) IEs are the total energy changes due to the atomic rearrangement of X_n impurities, as discussed in Section 2.2. As results, the absolute value of the RSCE error becomes less than ~ 0.1mRy per atom for both the ordered Pd_3X (X=Ru, Rh) alloys in $L1_2$ structure.

We now discuss the correction for each value of the n-body IEs (n=2~4). Tables 2, 3 and 4 show all the values for the 2-, 3-, and 4-body IEs with the sc-regions for the perturbed potentials up to the 1st- and 2nd-*nn* host atoms around X (=Ru, Rh) impurities in Pd, being used in the present RSCE (eq.(1)). For the 2-body IEs, we find that a certain degree of corrections (-0.10, -0.05, -0.06mRy) appear in the 2nd-*nn* IEs for X=Ru, Rh and the 6th-*nn* IE for X=Ru, while for the other IEs the absolute values for the corrections due to enlarging the sc-region for the perturbed potentials are less than ~ 0.03mRy. As results, the total correction from the 2-body IEs becomes as small as -1.55 and -0.46mRy for X=Ru and Rh (see Table 1).

Table 2. Calculated results (in mRy) for the 2-body IEs of (a) Ru_2 and (b) Rh_2 impurities in Pd, used for the RSCE of the total energies of Pd_3X (X=Ru, Rh) alloys. The perturbed potentials due to the insertion of impurities are redetermined up to the 2nd-*nn* host atoms around impurities in Pd. The values in the parentheses are the calculated results in which the perturbed potentials due to the insertion of impurities are redetermined up to the 1st-*nn* host atoms around impurities in Pd. The numbers of pairs per atom are also shown.

2-body	Number of pairs/atom	(a) 2-body IEs Ru_2 in Pd (mRy)	(b) 2-body IEs Rh_2 in Pd (mRy)
100 (2nd)	3	1.8604 (1.9641)	0.2056 (0.2580)
110 (4th)	6	2.1879 (2.2095)	0.1181 (0.1312)
111 (6th)	4	-0.2467 (-0.1860)	-0.0089 (0.0038)
200 (8th)	3	-0.0195 (-0.0353)	0.0111 (0.0095)
210 (10th)	12	0.0891 (0.0965)	0.0384 (0.0398)
211 (12th)	12	0.1012 (0.1142)	-0.0118 (-0.0092)
220 (15th)	6	-0.1198 (-0.1054)	-0.0132 (-0.0095)
221 (17th-a)	12	0.0443 (0.0617)	-0.0062 (-0.0029)
300 (17th-b)	3	-0.0398 (-0.0109)	-0.0157 (-0.0101)
310 (19th)	12	0.0028 (0.0268)	-0.0035 (0.0009)

For the 3-body IEs, as seen in Table 3, we find that the 3-body IEs for X=Rh are almost unchanged (the absolutes value for the corrections < ~ 0.002mRy). In contrast, the 3-body IEs for X=Ru change to a certain degree (~ -0.01mRy) for 224-, 228-, 2410, 2810, 448-3body IEs, while the corrections for the other 3-body IEs are fairly small (the absolute values of the corrections < ~ 0.004mRy). As results, the total correction from the 3-body IEs is as small as -1.26 and -0.05mRy for X=Ru and Rh (see Table 1).

For the 4-body IEs, as seen in Table 4, we find that the corrections due to enlarging the sc-region for the perturbed potentials are very small (the absolute values of the corrections < ~0.004mRy), except for the 222244- and 444444-4body IEs for X=Ru (0.01 and -0.02mRy). It is noted that the corrections of the 222244- and 444444-4body IEs for the total energy are small because the numbers of these clusters per atom are very small (3 and 2), as listed in Table 4. As results, the total correction from the 4-body IEs becomes very small (0.10 and 0.13mRy for X=Ru and Rh; see Table 1).

Table 3. The same as Table 2 for the 3-body IEs of (a) Ru_3 and (b) Rh_3 impurities in Pd (in mRy). The numbers of X_3 clusters per atom are also shown.

3-body	Number of X_3 clusters /atom	(a) 3-body IEs Ru_3 in Pd (mRy)	(b) 3-body IEs Rh_3 in Pd (mRy)
224	12	-0.1709 (-0.1609)	0.0277 (0.0269)
246	24	0.0580 (0.0604)	0.0224 (0.0231)
444	8	0.2541 (0.2540)	-0.0003 (-0.0027)
228	3	-0.3019 (-0.2886)	-0.0135 (-0.0138)
2410	24	-0.0938 (-0.0862)	-0.0019 (-0.0034)
2612	24	-0.0825 (-0.0786)	-0.0068 (-0.0062)
2810	24	-0.0731 (-0.0649)	-0.0139 (-0.0135)
21012	48	-0.0340 (-0.0316)	-0.0060 (-0.0052)
448	12	0.0308 (0.0395)	-0.0197 (-0.0205)
4412	24	0.0949 (0.0966)	-0.0031 (-0.0034)
4610	48	-0.0297 (-0.0264)	-0.0060 (-0.0055)
4812	24	-0.0113 (-0.0075)	-0.0002 (0.0002)
41010	24	0.0049 (0.0076)	0.0000 (0.0011)
668	12	-0.0438 (-0.0446)	-0.0060 (-0.0065)

Table 4. The same as Table 2 for the 4-body IEs of (a) Ru_4 and (b) Rh_4 impurities in Pd (in mRy). The numbers of X_4 clusters per atom are also shown.

4-body	Number of X_4 clusters/atom	(a) 4-body IEs Ru_4 in Pd (mRy)	(b) 4-body IEs Rh_4 in Pd (mRy)
222244	3	0.1713 (0.1622)	0.0296 (0.0289)
222444	8	0.0241 (0.0215)	0.0041 (0.0028)
222446	24	0.0156 (0.0147)	0.0048 (0.0005)
224446	24	0.0097 (0.0078)	-0.0064 (-0.0076)
224466	6	-0.0767 (-0.0794)	0.0048 (0.0067)
444444	2	0.0644 (0.0828)	0.0173 (0.0179)

4 Summary and Future Problem

We examined the delocalization of the perturbed potentials due to the insertion of X (=Ru, Rh) impurities in Pd by enlarging the sc-region for the perturbed potentials. We found that the discrepancy between the RSCE and the band-calculation results for the ordered Pd_3Ru alloy in $L1_2$ structure is satisfactorily corrected (from ~ 0.7mRy to ~ -0.1mRy per atom) by enlarging the sc-region for the perturbed potentials from the 1st-*nn* up to the 2nd-*nn* host atoms around Ru impurities in Pd. On the other hand, the correction for the RSCE error is small (from ~ 0.2mRy to ~ 0.1mRy per atom) for the ordered Pd_3Rh alloy in $L1_2$ structure, because the RSCE even with the sc-region for the perturbed potentials up to the 1st-*nn* host atoms around impurities in Pd already reproduces very well the band-calculation result for the ordered Pd_3Rh alloy in $L1_2$ structure, as seen in Tables 1-4. We clarified how the accuracy of the present RSCE depends on the sc-region for the perturbed potentials. As results, the absolute error of the RSCE becomes as small as ~ 0.1mRy for both the Pd_3X (X=Ru, Rh) alloys in $L1_2$ structure. It is obvious that, according to the requested accuracy for the calculated results, we must enlarge the sc-region for the perturbed potentials around impurities in Pd. However, if we need (for example) the accuracy of 1mRy per atom, it may be enough to redetermine self-consistently the perturbed potentials due to the insertion of impurities up to the 1st-*nn* host atoms around impurities in Pd.

We are now studying quantitatively the phase stability and phase diagram of the age-hardening Al-rich AlX (X=Mg, Cu, Zn) alloys by using the free energy calculations based on the present RSCE for the internal energies and the CVM for the configurational entropy, respectively, as shown in Refs. 11-13. It is also very interesting to study the effect of another element addition on the kinetics of age-hardening behaviors of Al-rich multi-element alloys, such as Cu in Al-rich AlMgSi alloys [18, 19]. We plan to develop the calculation program for Monte Carlo simulations [20-22] which may be useful as one of material design tools for the high quality of age-hardenable Al-rich multi-element alloys. We believe Al-rich multi-element alloys offer a good example how thermodynamical calculations can reveal quantitative insight into the underlying age hardening behavior.

As another application of the present RSCE approach for the internal energy, it may be interesting to study the morphology in the nano-scale spinodal decomposition phase [23,24], such as the magnetic network shown in Fig.5 in Ref. 24, which occur in the quenching and annealing processes in the dilute magnetic semiconductors such as (Ga, Mn) N and (Ga, Mn) As. At present we can take into account the lattice distortion effect for the chemical and magnetic IEs derived from a dilute limit [25, 26], which may be important for Mn impurities (at Ga sites) in the GaAs and GaN semiconductors because the difference between the ionic radii (2.67 and 2.38 a.u.) of Ga and Mn atoms in fcc structure is considerably large [27]. Morphological features expected by Monte Calro simulations with such energetics will be reported in the future.

Acknowledgement

The authors are grateful for the financial support from the Ministry of the Education, Culture, Science and Technology (JSPS KAKENHI Grant Nos, 15K06422 and 16K06710). This study was partly supported by New Energy and Industrial Technology Development Organization of Japan (NEDO) Grant (P16010). One of the authors (T. Mohri) appreciates their support. This study was almost completed when T. Hoshino was at the Institute for Material Research, Tohoku University. C. Liu and T. Hoshino would like to express their sincere thanks to the crew of the Center for Computational Materials Science of the Institute for Materials Research, Tohoku University, for their continuous support of the SR16000 supercomputing facilities.

References

1 M. Asato, T. Mizuno, T. Hoshino, and H.Sawada, Mater. Trans. **42** (2001), 2216-2224.
2. T. Hoshino, M. Asato, R. Zeller, and P. H. Dederichs, Phys. Rev. B**70** (2004), 094118(1-7)
3. T. Hoshino, M. Asato, S. Tanaka, F. Nakamura and N. Fujima, Intermetallics **14** (2006), 913-916.
4. F. Nakamura, T. Hoshino, S. Tanaka, K. Hirose, S. Hirosawa, and T. Sato,Trans. Mat. Res. Soc. Japan **30** (2005), 873-876.
5. M. Asato, H. Takahashi, T. Inagaki, N. Fujima, R. Tamura, and T. Hoshino, Mater. Tran. **48** (2007), 1711-1716.
6. T. Hoshino, W. Schweika, R. Zeller, and P. H. Dederichs, Phys. Rev. B **47** (1993), 5106-5117.
7. T. Hoshino, M. Asato, N. Fujima, J. Alloy. Compd. **504S** (2010), S534-S537.
8. M. Asato, A.Settles, T. Hoshino, T. Asada, S. Blügel, R. Zeller, and P. H. Dederichs, Phys. Rev. B **60** (1999), 5202-5510.

9. T. Hoshino, M. Asato, T. Mizuno, and H. Fukushima, Mater. Trans. **42** (2001), 2206-2215.
10. T. B. Massalski, H. Okamoto, P. R. Subramanian, and L. Kacprazak: *Binary Alloys Phase Diagrams*, 2nd. ed., (ASM International, New York , 1990.
11. C. Liu, M. Asato, N. Fujima, T. Hoshino, Y. Chen, and T.Mohri, Mater. Trans. **59** (2018), 338-347.
12. C. Liu, M. Asato, N. Fujima, T. Hoshino, Y. Chen, and T.Mohri, Mater. Trans. **59** (2018), 883-889.
13. M. Asato, C. Liu, N. Fujima, T. Hoshino, Y. Chen, and T.Mohri, World Academy of Science, Engineering and Technology International Journal of Materials and Metallurgical Engineering Vol. **12**, No.3, 119-123 (2018).
14. C. Liu, M. Asato, N. Fujima, T. Hoshino, Y. Chen, and T. Mohri, Mater. Trans. **59** (2018), 1669-1676.
15. R. Zeller, Phys. Rev. B**55** (1997), 9400-9408.
16. R. Zeller, M. Asato, T. Hoshino, J. Zalbloudil, P. Weinberger, and P. H. Dederichs, Philos. Mag. B **78** (1998), 417-422.
17. K. Wildberger, P. Lang, R. Zeller, and P. H. Dederichs, Phys. Rev. B **52** (1995), 11502-11508.
18. S. Hirosawa, F. Nakamura, T. Sato, and T. Hoshino, J. Jpn. Inst. Light Met. **56** (2006), 621-628.
19. J. Kim, F. Kobayashi, and T. Sato : Mater. Trans. **52** (2011), 906-913.
20. T. Hoshino, R.Zeller, P.H. Dederichs and T. Asada, Computational Physics as a New Frontier in Condensed Material Research, Ed. By H. Takayama *et al*, Physical Society of Japan, Tokyo, (1995), pp. 105-113.
21. W. Schweika, *Disordered Alloys: Diffuse Scattering and Monte Carlo Simulations*, (Springer Tracts in Modern Physics **141**, 1998), pp. 51-57.
22. S. Hirosawa and T. Sato : J. Jpn. Inst. Light Met. **54** (2004), 121-127.
23. K. Sato, H. Katayam-Yoshida, and P. H. Dederichs : Jpn. J. Appl. Phys. **44** (2005), L948-L951.
24. T. Diel, K. Sato, T. Fukushima, A. Bonnani, M. Jamet, A. Barski, M .Tanaka, P. H. Hai, and H. Katayama-Yoshida, Rev. Mod. Phys. **87** (2015), 1311-1377.
25. P. Lang, L. Nordstrom, K. Wildberger, R. Zeller, P. H. Dederichs, and T. Hoshino, Phys. Rev. B **53** (1996), 9092-9107.
26. T. Hoshino, N. Fujima, M. Asato, and H. Tatsuoka : J. Alloys Compd. **504** (2010), S531-S533.
27. C. Kittel : *Introduction to Solid State Physics*, 7th ed., (Jhon & Wiley Sons. Inc. New York, 1996), see Table 9.

Disintegration and Recycling of Multi-layered Glass Fiber Reinforced Polymer Composites via Superheated Steam

Chi Hoong Chan[1], Minato Wakisaka[1] and Haruo Nishida[1, a]

[1]*Graduate School of Life Science and Systems Engineering, Kyushu Institute of Technology, 2-4 Hibikino, Wakamatsu-ku, Kitakyushu, 808-0196 Fukuoka, Japan.*

Abstract. To increase viability of recycling of robust and large-scale multi-layered glass fiber reinforced polymer composites (GFRPs), disintegration of the multi-layered GFRPs and recovery of fibers/resin-derived materials were investigated using superheated steam (SHS) under normal pressure where a very quick heat transfer to GFRP was possible. The SHS treatment of 4 different types of multi-layered GFRP products were conducted at 350 °C in an oxygen-free environment up to 3 hours. The SHS treated GFRPs were easily disintegrated into each layer. The separated layers were divided into components: glass fibers, oligomers, and inorganic fillers after subsequent thermal and ultrasonication processes. Finally, clear glass fibers were recovered, and matrix resin was also recovered as soluble oligomers consisting of phthalates, glycols, and styrene units due to partial chain cleavage of cured resin. These results clearly showed the viability for the recycling of actual large-scale multi-layered GFRP products.

1 Introduction

Thermosetting polymers have been proved to be more suited in applications for harsher environments because of their three-dimensional network structures that made the material stronger, more stable at high temperature, and more resistant towards chemical stimulations compared to thermoplastics. In addition, by adding high strength fibers to thermosets, high-performance fiber-reinforced polymer composites (FRPs) can be obtained. Various high-strength / high-modulus fibers such as aramid fibers, carbon fibers, glass fibers, etc. have been used extensively to reinforce thermosets [1]. The composites are tougher, more durable and lighter in weight compared to traditional materials like steel [2]. These benefits are contributable to higher ease in handling and longer service life of products. Therefore, FRPs have been using extensively in various industries such as civil engineering [3], aerospace [4], marine [5], and automotive [6,7].

However, the durability of FRPs has been called into questions as they are nearing their expected service life, due to their lack of a proper life-cycle assessment [2]. Conventional thermoset FRPs cannot enter reprocessing or recycling systems which consist of various raw materials including, fibers, sizing agents, fillers, and other additives, making their recycling very difficult. For this reason, FRPs such as glass fiber reinforced polymer composite (GFRP) have been usually disposed via land-filling or incineration [8]. In Japan, the waste disposal using land filling method is secondary compared to incineration as the primary mode of disposal [9]. The incineration of wastes allows the energy recovery from their organic components. However, incombustible materials from the incineration process are usually collected, treated, and finally disposed by land-filling. Therefore, secondary usage of FRP-derived materials should be encouraged.

FRP-recycling mainly focused on recovering the more expensive reinforcing fibers, e.g. pyrolysis [10], with oxide semiconductors [11] and subcritical fluids [12], while little attention has been paid to polymer components to recover, for example, a styrene-fumaric acid copolymer [13,14]. A recent review by Oliveux et al. classified the FRP recycling into 3 different modes, namely, mechanical, thermal, and solvolysis [15]. Solvolysis method that increases the economic viability of FRP recycling by allowing the total recovery of fibers and polymers has been demonstrated. Recently, hydrothermal treatment with steam of GFRP has been investigating without using any chemical agents. Shi and Bao reported the superheated steam (SHS) degradation of a cured unsaturated polyester (UP) and the single layered GFRP prepared by hand lay-up molding. The GFRP was treated by SHS at 340 – 450 °C to remove resins and recover glass fibers with char attached [16]. Nishida et al. also treated the single-layered GFRP with a gel-coated surface layer by SHS at 300 °C to remove resins and recover glass fibers [17]. Despite the many previous studies, there are a few reports [18] published regarding the recycling of actual multi-layered GFRP products, which have been used as large-scale structures for buildings and civil engineering constructions. In this study, the disintegration viability of multi-layered GFRP

[a] Corresponding author: nishida@lsse.kyutech.ac.jp

was investigated to facilitate the recovery of glass fibers and organic components from the robust multi-layered GFRPs using SHS under normal pressure. Four types of robust GFRP products with different layered structures were used in SHS treatments. As results, the GFRPs were successively separated to each layer while hydrolyzed oligomers were also extracted. With subsequent thermal degradations, clear glass fibers were also recovered.

2 Materials and Methods

2.1 Materials

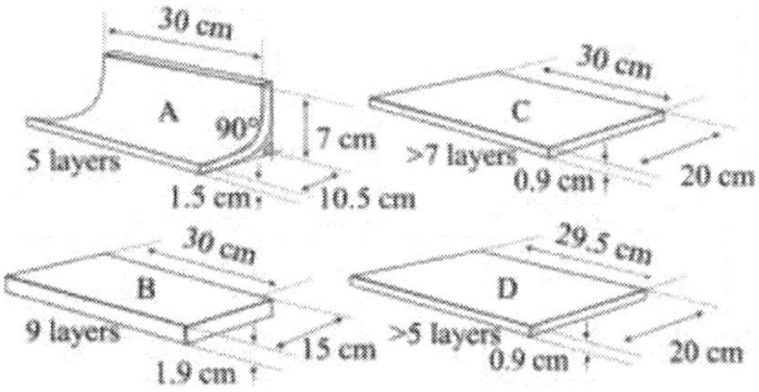

Fig. 1 Dimensions and layer structures of GFRP samples.

Four different types of multi-layered GFRPs with various dimensions, shapes, and layer structures, which were listed in Table 1 and Fig. 1, a typical unsaturated polyester (UP, poly (phthalic acids-fumaric acid-glycols) including styrene monomer of 32.8 wt%), and initiator: methyl ethyl ketone peroxide (MEKP) were provided from Kurimoto, Ltd. (Osaka, Japan). Five kinds of solvents: chloroform, n-hexane, acetone, toluene, and methyl alcohol were purchased from Wako Pure Chemical Industries, Ltd. and used as received.

2.2 Methods

2.2.1 Curing of Unsaturated Polyester

Unsaturated polyester (UP) was cured by polymerization of styrene using MEKP at 1.0 vol% according to the preparative procedure recommended by the producer. The UP including MEKP was stirred at 25 °C for 24 h, followed by heating at 70 °C for 3 h, resulting in hardening of UP to cured resin.

2.2.1 SHS and Thermal Treatments Procedures

The multi-layered GFRPs (Table 1 and Fig. 1) and UP were treated by SHS in a SHS oven QF-5200C (Naomoto Corp., Osaka, Japan) at 350 °C for 1 or 3 h. After the SHS treatment, GFRPs were immediately cooled to ambient temperature using a fan and weighed. The SHS-treated samples were easily disintegrated to each layer even by handwork. Mechanically separated layers were then subjected to thermal treatment in a pre-heated furnace MMF-1 (AS ONE Corporation, Osaka, Japan) at 500 °C for 1 h under atmospheric conditions. The heat-

Table 1. Multi-layered GFRPs.

Samp.	Thickn. (mm)	Layer Structure	
		Number	Components
A	15	5	glass fiber inner core layer; strands; and outermost mat layers at both sides
B	19	9	mortar inner core (silica); glass fiber, 3-layers mats; and outermost single layer are with fillers
C	9	>7	mortar inner core (silica); glass fiber, mat layer; and outermost intercalated 2-layers
D	9	>5	mortar inner core (silica) layer; glass fiber, 2-layers mats; outermost layer with chopped glass fibers

treated sample layers were then allowed to cool down before subjecting to ultrasonication in an US-103 (Sansyo Co., Ltd., Tokyo, Japan) to remove attached organic materials. Finally, glass fibers were recovered after oven-drying at 105 °C for 24 h. For comparison sake, GFRP sample A in Table 1 was also heat-treated at 350 °C for 3 h without steam under atmospheric conditions. After the heat-treatment, sample A was immediately cooled to ambient temperature using a fan and weighed.

2.2 Characterizations

Chemical structures of GFRP samples were analyzed with a Fourier transform infra-red (FT-IR) spectrometer (Nicolet iZ10, ThermoScientific, Waltham, MA, USA) via a transmission mode using KBr pellet. The spectrum was recorded in a wavenumber range of 400 to 4000 cm^{-1} with spectral resolution of 4 cm^{-1} and 16 scans per sample. The surface morphologies of SHS treated GFRPs were observed under a 3D laser scanning confocal microscope (LSCM) model VK-X 100/105 (KEYENCE Corporation, Osaka, Japan) under prescribed conditions of laser: red semiconductor laser, λ=658 nm, 0.95 mW, and pulse width of 1 ns using a depth composition procedure to check the presence of residual materials and quality of glass fiber surfaces. Proton (^1H) nuclear magnetic resonance (NMR) spectra of chloroform-extracted materials from SHS-treated GFRPs were recorded on a 500 MHz JNM-ECP 500 FT-NMR system (JEOL Ltd., Tokyo, Japan). Tetrachloroethane-d2 (TCE-d2) was used as a solvent. Size-exclusion chromatography (SEC) measurements of chloroform-extracted materials from SHS-treated GFRPs were analysed using TOSOH HLC-8320 GPC system (Tosoh Corporation, Tokyo, Japan) with refractive index (RI) and ultra-violet (UV) detectors at 40 °C using TOSOH TSK Gel Super HM-M column (linearity range: 1×10^3–8×10^6; molecular weight (MW) exclusion limit 4×10^8) and a chloroform eluent (0.6 mL/min). Calibration curves for SEC analysis were obtained using polystyrene standards with low polydispersity and number-average molecular weight (Mn) of 5.89×10^2, 1.01×10^3, 2.5×10^3, 5.43×10^3, 9.49

× 10^3, 1.37 × 10^4, 3.72 × 10^4, 9.89 ×10^4, 1.89 × 10^5, 3.97 × 10^5, 7.07 × 10^5, and 1.11 × 10^6. Approximately 10 mg of sample was dissolved in 2 mL chloroform, stirred for 2 h at room temperature, and filtered through a membrane filter with 0.45 μm pore size to be suitable for use in the SEC analysis.

3 Results and Discussions

3.1 Morphological and Weight Changes after SHS and After Thermal Treatment

To disintegrate the multi-layered GFRP samples and recover clear glass fibers, SHS-treatment at 350 °C for 3 h, thermal degradation at 500 °C for 1 h, and subsequent ultrasonication were conducted. Weight loss values during the successive treatments were listed in Table 2. After 3 h of treatment, all GFRP samples decreased in weight by 13.2–25.8 %, and easily disintegrated into each layer even by handwork (Fig. 2). After 1 h of the SHS treatment, the GFRP samples were disintegrable by weak mechanical stress. Robust multi-layered GFRPs of 9–19 mm in thickness (Table 1), can be penetrated by SHS at 350 °C from outermost layers of GFRPs and diffuse into inner core layers to disintegrate the layer structures. As a comparison, s imple heat treatment of GFRP sample A was also conducted in the same oven at 350 °C for 3 h without SHS. After the heat treatment, sample A partially changed color into pale brown and decreased weight by 3.24 % but did not disintegrate. Other GFRP samples showed the same changes. These results clearly show the significant effect of SHS treatment to disintegrate the robust multi-layered GFRP products.

The SHS-treated GFRP samples were subjected to subsequent thermal degradation at 500 °C for 1 h with each layer. After the thermal degradation, the layers showed additional weight loss by 2.1–24.7 %, varying dependent on layer components. Particularly, the outermost layers showed the highest weight loss, indicating that the outermost layers have the highest number of organic components, which remained even after the SHS treatment. While the inner core layers showed the lowest weight loss, because the layers are consisted of a large amount of fillers such as SiO_2. As shown in Fig. 2, GFRP samples and separated layers after 3 h of SHS treatment changed color from light gray to brown / dark gray, which was attributed to the steam-degraded resin (SDR) derived from cured UP. Some layers clearly showed embedded glass fibers (Fig. 2) and the inner core layers showed fragmentation of mortar layer.

The thermally treated layers were subsequently subjected to ultrasonication to separate the remaining SDR on glass fiber surfaces in distilled water, resulting in clean glass fibers and recovery of powdery SiO_2 fillers from outer layers and inner mortar core layers (Fig. 3). Although the thermal and ultrasonication treatments were used to achieve the recovery of clear fibers, it was also

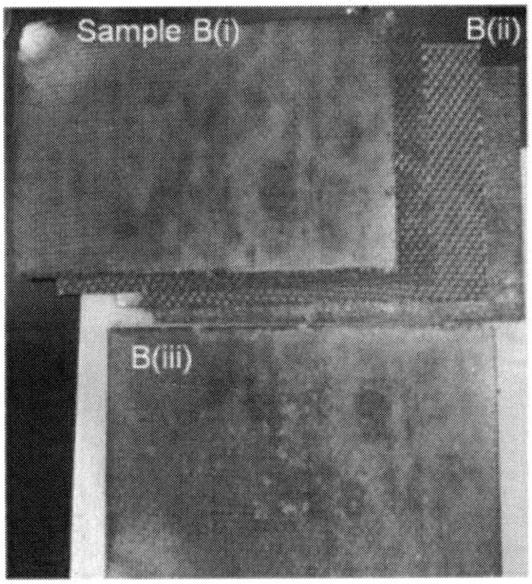

Figure 2. Separated GFRP sample layers after SHS treatment at 350 °C for 3 h. (a) sample B: (i) outermost, (ii) multiple glass fiber mats and (iii) mortar core layers.

Table 2 Weight loss values of multi-layered GFRPs after SHS and thermal treatments

Samp.	Initial Mass (kg)	SHS (wt.%) [a]	Thermal Treatment [b] Separated Components	(wt%)
A	1.32	25.8	outermost layers	24.7
			glass fiber strand layers	11.6
			glass fiber inner core	8.3
B	2.42	13.2	outermost layers	4.5
			glass fiber mat layers	3.6
			mortar-based inner core	3.1
C	2.52	15.9	outermost layers	15.8
			glass fiber strand layers	11.3
			mortar-based inner core	5.5
D	1.28	15.6	outermost layers	5.2
			mortar-based inner core	2.1

[a] SHS treatment at 350 °C for 3 h
[b] Thermal treatment in a furnace at 500 °C for 1 h

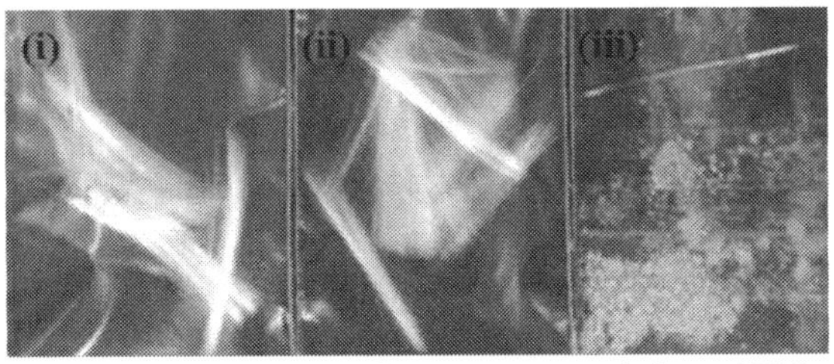

Figure 3. Glass fibers and mortar recovered from GFRP sample C after thermal treatment at 500 °C for 1 h (i) Outermost, (ii) glass fiber mat, and (iii) mortar core origins.

confirmed that these post-treatments were not necessarily needed for the recycling. It was found that other simpler methods such as vibration on a screen were also usable for separating SDR attached on the fiber surfaces. Morphologies of the recovered fibers and chemical structures of SDRs were analyzed respectively.

3.2. Laser Scanning Confocal Microscopy

SHS treated GFRP samples and its subsequent thermal and ultrasonication treatments were observed under LSCM using a depth composition procedure by stitching images together at different focal length and magnification. As shown in Fig. 4 (a-d), glass fibers were clearly seen on the surfaces of the disintegrated layers by SHS treatment for 3 h. The matrix polymer suffered noticeable degradation during the SHS treatment. Although the outlines of glass fibers were visible for all the samples, brown and black-colored SDR particulates were also observed attached on fiber surfaces. After the subsequent thermal treatment at 500 °C for 1 h, all the samples turned white in color from the white particulates on fiber surfaces, which were determined from FT-IR analysis to be calcium carbonate ($CaCO_3$) filler. After the ultrasonication and washing, the white particulates were completely removed. Finally, clear glass fibers without any accretions were recovered as shown in Fig. 4(iii). During the SHS treatment, the degradation of GFRPs must start by steam penetration from outermost layers then diffuse to the inner layers, resulting in successful disintegration into each layer. However, due to the thickness and firm networks of layered glass fibers, generated SDRs were trapped inside the composite layers without being carried away by steam flow.

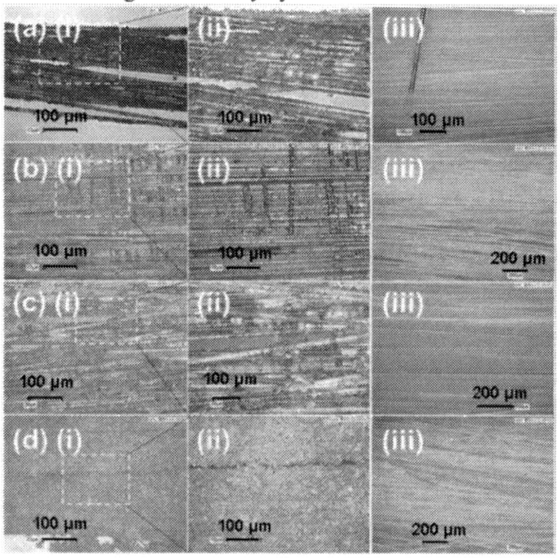

Figure 4. LSCM images of GFRP samples after SHS treatment at 350 °C for 3 h, and their enlarged view (a)(i) sample A, (b)(i) sample B, (c)(i) sample C, (d)(i) sample D, in which (ii) enlarged samples and (iii) after thermal treatment respectively.

3.3 FT-IR

After the SHS treatment, SDRs were recovered from the separated layers. However, each layer consists of a distinct composition of raw materials, showing different decomposition behaviors. Cured UP without glass fiber was used as a control. From FT-IR spectra shown in Fig. 5, all samples showed the characteristic triple-peaks νaromatic C-H at 3083, 3060 and 3030 cm^{-1} and

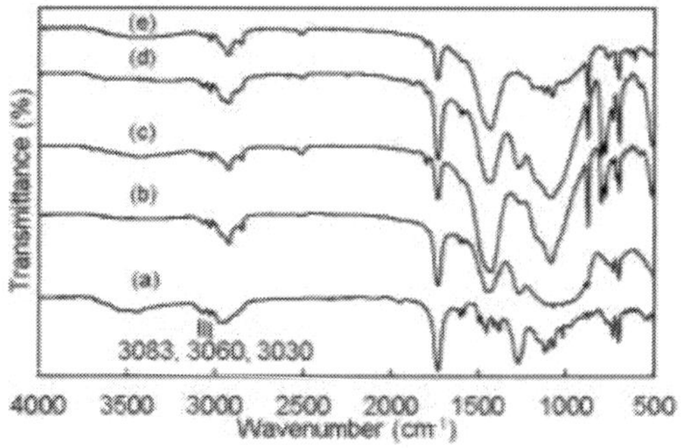

Figure 5. FT-IR spectra of (a) UP after curing and pristine GFRPs: (b) sample A, (c) B, (d) C and (e) D

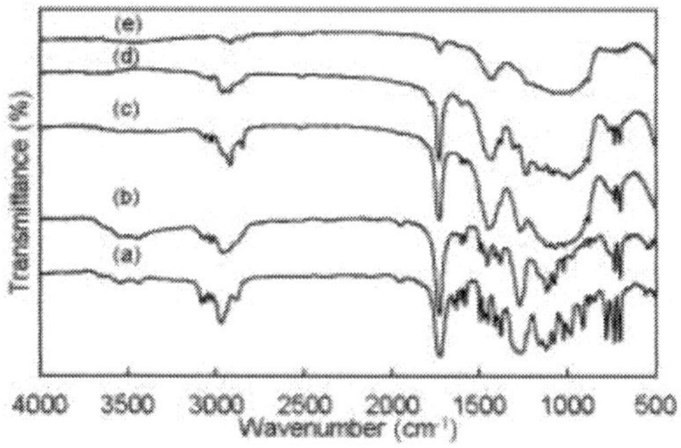

Figure 6. Changes in FT-IR spectrum of UP and GFRP (sample A) in during treatments, (a) pristine UP, (b) cured UP at 70 °C for 3 h, (c) sample A, and after (d) 1 h, (e) 3 h of SHS treatment at 350 °C, and (f) thermal decomposition at 500 °C for 1 h and ultrasonication

δaromatic C-H, out-of-plane at 700 cm^{-1}. Other characteristic peaks νC=O and νC-O of ester groups were also found at 1730 and 1200-1300 cm-1. Whereas, νC=C peak at 1630 cm^{-1} derived from styrene monomer cannot be found in all spectra, indicating that all polymer matrices were cured completely.

Sample A was chosen as a representative GFRP sample to show changes in FT-IR spectrum with successive treatments, i.e., SHS treatment, thermal treatment, and ultrasonication were shown in Fig. 6. The characteristic triple-peaks νaromatic C-H were found in cured UPs, pristine and SHS (1 h) treated GFRPs as shown in Fig. 6(a–d). After 3 h of SHS treatment, the characteristic peaks almost disappeared except for νC=O at 1730 cm^{-1} of ester groups (Fig. 6(e)). However, the νC=O peak was not detected after the thermal treatment and ultrasonication (Fig. 6(f)). It has been reported that characteristic peaks νC-O of aryl esters are identified to be in the range of 1210–1290 cm-1 [21]. The maximum absorption peak of νC-O was found at 1200-1300 cm^{-1}, which was also observed in pristine UP and all GFRPs as derived from tere/iso/ortho-phthalates (Fig. 5). The intensity of νC-O peak diminished and subsequently disappeared after the thermal treatment and ultrasonication (Fig. 6(f)). Finally, a simple FT-IR spectrum was obtained, one broad peak and multiple peaks in ranges of 830-1350 and 400-520 cm^{-1} for νSi-O-Si asymmetric and νSi-O-Si,symmetric, respectively, mostly derived from the glass fiber. Based on the results of FT-IR analysis matrix resin, it was clearly observed that SHS at 350 °C effectively degraded the cured UP by cleaving not only ester bonds, but also C-C bonds as observed as diminishing in many characteristic peaks, and SDR was completely removed by the subsequent thermal treatment and ultrasonication. Finally, the clear glass fibers were recovered.

3.4 Analysis of SDR (^{1}H-NMR and SEC)

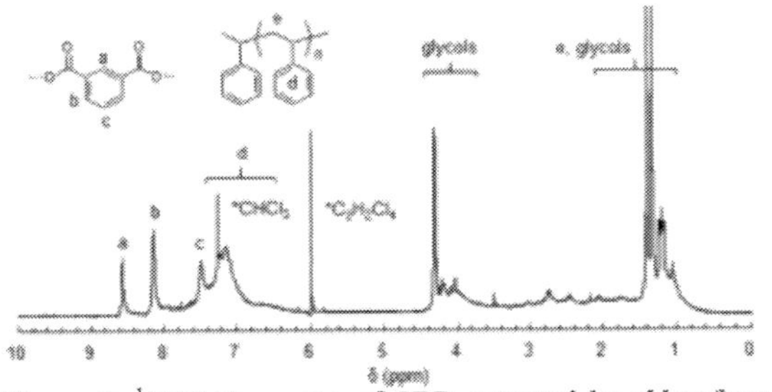

Figure 7. ^{1}H-NMR spectra of SDR extracted by chloroform from sample A after 1 h of SHS treatment at 350 °C

Previously, the hydrolytic degradation of cured UP has been conducted via subcritical water at 230 °C and 2.8 MPa with NaOH or KOH, resulting in recovery of styrene-fumaric acid copolymers [13]. In this study, milder and more feasible SHS conditions: 350 °C under normal pressure were selected to recycle large scale products. After the SHS treatment for 1 h, SDR from GFRP sample A was extracted by chloroform and analyzed with 1H-NMR using TCE-d2 as a solvent. Fig. 7 shows the 1H-NMR spectrum of SDR extracted by $CHCl_3$. Main components of SDR were identified as a mixture of isophthalate, glycols, and styrene moieties detected in chemical shift ranges of 7.5-8.7, 3.5-4.5 / 0.6-1.5, and 6.2–7.4 / 1.0-2.0 ppm, respectively.

The SDR extracted from GFRP sample A from the same sample in Fi.g 7, was a low molecular weight oligomer. Number-average (M_n) and weight-average molecular weight (M_w) values of 1.2×10^3 and 5.7×10^3, were obtained using RI detector. When used UV detector, M_n and M_w values were estimated as 0.7×10^3 and 4.7×10^3, respectively. The polydispersity index (PDI) values were determined are quite wide, 4.81 and 6.50 for RI and UV measurements, respectively. The differences between RI and UV data would be attributed to heterogeneous distribution of components having aromatic moieties such as styrene unit sequence in the lower MW range. Overall, the SDR was found to be an oligomer with high PDI values with various main and side chain-length. consist of styrene/fumarate-isophthalate-glycol copolymers.

4 Conclusion

In this study, the viability of disintegrating and recycling robust large-scale multi-layered GFRP products using SHS was investigated. Four types of multi-layered GFRPs of different layer structures were employed as actual product samples. GFRPs were easily disintegrated and separated to each layer via SHS treatment at 350 °C under normal pressure. The separated layers were easily separated for subsequent recycling processes. During the SHS treatment, UP matrix was degraded partially and escaped via exhaust. Remaining SDR was attached on glass fiber surfaces as particulates. Subsequent thermal treatment at 500 °C and ultrasonication separated SDR from glass fiber surface, resulting in recovery of clear glass fibers. From FT-IR and ^{1}H-NMR analyses of SDR, it was confirmed that the SDR was soluble in organic solvent and a low MW oligomer, consisting of phthalates, glycols, and styrene units due to partial cleavage of cured UP chains. Therefore, it was confirmed that the sequential disintegration of multi-layers and recycling system consisting of SHS, thermal, and ultrasonication treatments was successful in recovering clear glass fibers as well as soluble oligomers from actual large-scale multi-layered GFRP products.

References

[1] P.K. Mallick. *Fiber-Reinforced Composites: Materials, Manufacturing, and Design. Third Edition.* (CRC Press, Boca Raton, 2007).
[2] L.S. Lee, R. Jain, Clean Technol. Envir. **11**:247-249 (2009).
[3] L.C. Hollaway, Constr. Build. Mater. **24**:2419-2445 (2010).
[4] F.C. Robertson, Br. Polym. J. **20**:417-429, (1988).
[5] A.P. Mouritz, E. Gellert, P. Burchill, K. Challis, Composite Structures **53**:21-42, (2001).
[6] P. Beardmore, Compos. Struct. **5**:163-176, (1986)
[7] G. Savage, Failure Anal. **17**:92-115, (2010)
[8] M.C.S. Ribeira, A.C. Meira-Castro, F.G. Silva, J. Santos, J. P. Meixedo, A. Fiúza, M.L. Dinis, M.R. Alvim, Resour., Conserv. and Recycl. **104**:417-426, (2015).
[9] N. Tanaka, Y. Tojo, T. Matsuto, J Mater Cycles Waste Manag **7**:104-111, (2005).
[10] G. Grause, T. Mochizuki, T. Kameda, T. Yoshioka. J Mater Cycles Waste Manag **15**:122-128, (2013).
[11] H. Shima, H. Takahashi, J. Mizuguchi, Mater. Trans. **52**:1327-1329, (2011)
[12] T. Iwaya, S. Tokuno, M. Sasaki, M. Goto, K. Shibata, J. Mater. Sci. **43**:2452–2456, (2008).
[13] T. Nakagawa, M. Goto, Polym. Degrad. Stab. **115**:16-23, (2015)
[14] K. Maekawa, K. Shibata, M. Iwai, A. Endou. Hitachi Chemical Technical Report No. **42**: 21-24, (2004)
[15] G. Oliveux, L.O. Dandy, G.A. Leeke, Prog. Mater. Sci. **72**:61-99, (2015)
[16] J. Shi, L. Bao, Jpn. J. Appl. Phys. **50**: 01AJ01-1-5, (2011).
[17] H. Nishida, N. Mouri, K. Oshima, T. Yamamoto, Reinf. Plas. **60**:13-20, (2014).
[18] K. Hamada, K. Imai, H. Uematu, H. Kawakita Kochi Koshi Houkoku **21**:8-12, (1990).
[19] Y.-J. Huang, J.-S. Leu, Polymer **34**:295-304, (1993).
[20] J.C. Yang, M.J. Jablonsk, J.W. Mays, Polymer **43**:5125-5132, (2002).
[21] A.H. Kuptsov, G.N. Zhizhin, *Handbook of Fourier Transform Raman and Infrared Spectra of Polymers, Volume 45* (Elsevier Science, Amsterdam, 1998).

Preparation of hydrophobic nanofibers by electrospinning of PMMA dissolved in 2-propanol and water

Hui-Yi Chang, Chao-Ching Chang[a], Liao-Ping Cheng

Department of Chemical and Materials Engineering, Tamkang University, New Taipei City, Taiwan, ROC

Abstract. In this study, we adopted rubbing alcohol (2-propanol/water = 7.8/2) as the solvent to prepare hydrophobic poly(methyl methacrylate) (PMMA) nanofibers (with submicron scale diameters) by electrospinning. In the literature, the general solvents, such as acetone, tetrahydrofuran, chloroform, toluene, etc., to dissolve PMMA are harmful and not environmentally friendly. 2-Propanol and water are both not hazardous to humans and the environment. PMMA dissolved in rubbing alcohol can be electrospun near room temperature. The solutions were heated at ~60 °C and allowed to cool to room temperature. Controlling the solution concentration and electrospinning parameters, fibers with diameters of 0.65~0.85 μm were obtained. The electrospun PMMA mats were hydrophobic with contact angles > 130 ° and showed good water resistance.

1. Introduction

Electrospinning is a technique which draws polymer solutions or polymer melts to submicro- or nanofibers. Due to the advantages of spun fibers, such as high porosity and high ratio of surface area to volume, electrospinning has shown significant potential for tissue engineering, drug delivery, filtration, sensors, fuel cells, photoelectric devices, and other applications [1,2]. For environment-friendly processes and stable long-time-use products, it is necessary to consider seriously the polymer and solvent used for fiber spinning. Water-soluble and hydrophilic polymers can be dissolved in water, alcohols or their mixtures, but the spun fibers do not possess moisture resistance. Thus they are limited to temporary use and then removed in water. After absorbing water, the fibers will reduce or lose their functions due to the structural deformation, even gradually dissolve in the water.

[a]Corresponding author: ccchang@tku.edu.tw

Poly(methyl methacrylate) (PMMA) is a hydrophobic and common material. PMMA submicro- and nanofibers prepared by electrospinning have been studied and shown potential for tissue engineering and biomaterials [3–6]. The common solvents, such as acetone, tetrahydrofuran, chloroform, toluene, etc., to dissolve PMMA are harmful and not environmentally friendly. The solubility of hydrophobic polymers in water or alcohols is often very low. However, 2-propanol mixed with water, i.e., rubbing alcohol, dissolves some hydrophobic polymers, including PMMA. 2-Propanol and water are not hazardous to humans and the environment. It has been known that PMMA can be dissolved in rubbing alcohol with particular ratios of 2-propanol and water [7]. In this study, PMMA–rubbing alcohol solutions were prepared and electrospun. Controlling the PMMA concentration and electrospinning parameters, PMMA fibers with diameters of 0.65~0.85 μm were obtained. The electrospun mats were hydrophobic with contact angles > 130 ° and showed good resistance.

2. Experimental Section

2.1 Materials

Poly(methyl methacrylate) (PMMA) was purchased from Sigma-Aldrich (weight-average molar weight of 120,000 g/mol) and Alfa-aesar (weight-average molar weight of 400,000 g/mol). 2-Propanol with purity over 99.5% was purchased from J.T. Baker. Deionized water and distilled water were obtained from laboratory. All chemicals were used as received.

2.2 Preparation of PMMA fibers

The polymer solution was prepared by dissolving PMMA in rubbing alcohol (2-propanol : water = 7.8 : 2, w/w), stirring at ~60 °C for 1 hour, and then cooling to room temperature. A vertical electrospinning setup (Falco Tech Enterprise Co., Ltd, Taiwan) was used. The PMMA solution was in a 5 mL plastic syringe, and a syringe pump (New Era Pump Systems, Inc.) was used to promote the solution. The syringe was connected to a needle (inner diameter of 0.42 mm) through a Teflon capillary tube. A stainless steel plate covered with aluminum foil was employed as collector (12 cm below the needle). An electric field between the needle and collector was generated using a high voltage power supply. The temperature was maintained at 30 °C and the relative humidity was about 45%.

2.3 Fiber analysis

The sub-microfibers were observed by a field emission scanning electron microscope (FE-SEM, Hitachi S-4800), and the diameter of the fibers was calculated by using software Image J based on the SEM images. The contact angle of water on the mat surface was measured by a contact angle goniometer (FTA-125, First Ten Angstroms) at room temperature. A 3 μL drop of water was placed onto the mat surface. The image was taken and the contact angle was measured by means of shape analysis of the sessile drop.

3. Results and Discussion

Shenoy *et al.* [8,9] proposed a semi-empirical formula [equation (1)] to predict the result of electrospinning. The solution entanglement number $(n_e)_{soln}$ is defined as the ratio of the weight–average molecular weight (M_w) to the entanglement molecular weight in solution ($(M_e)_{soln}$). The latter can be calculated by means of the polymer volume fraction (ϕ_p) and the entanglement molecular weight in melt (M_e).

$$(n_e)_{soln} = \frac{M_w}{(M_e)_{soln}} = \frac{(\emptyset_p M_w)}{M_e} \quad (1)$$

$$\emptyset_p = \frac{\frac{polymer\ weight\ (g)}{polymer\ density\ (g/cm^3)}}{\frac{polymer\ weight\ (g)}{polymer\ density\ (g/cm^3)} + \frac{solvent\ weight\ (g)}{solvent\ density(g/cm^3)}} \quad (2)$$

When $n_e < 2$, only beaded morphology is predicted; when $n_e > 3.5$, complete fiber formation is predicted. On the basis of equation (1), complete PMMA fibers will appear at concentrations of 42 wt% and 13.5 wt% for PMMA with M_w~120,000 and M_w~400,000, respectively (Figure 1). However, in this study the experiment results did not agree with such calculated results.

https://doi.org/10.1051/matecconf/201926403004

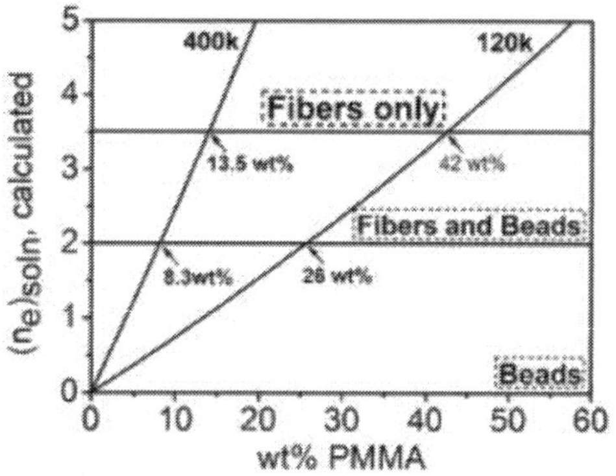

Figure 1. The calculated solution entanglement number $(n_e)_{soln}$ as a function of concentration for PMMA in IPA/water.

Solutions of PMMA with M_w~120,000 volatilized too rapidly to be electrospun successfully. All beads were observed at 5 wt%, and the beads and fibers was observed at 8 wt% (Figure 2). Solutions gelled at the tip of the needle quickly at higher concentrations. Beads and fibers also appeared at 1 wt% of PMMA with M_w~400,000 (Figure 2). However, good beads-free fibers appeared when the concentration reached 1.5 wt%. The fibers were straight and the mat was uniform (Figure 3).

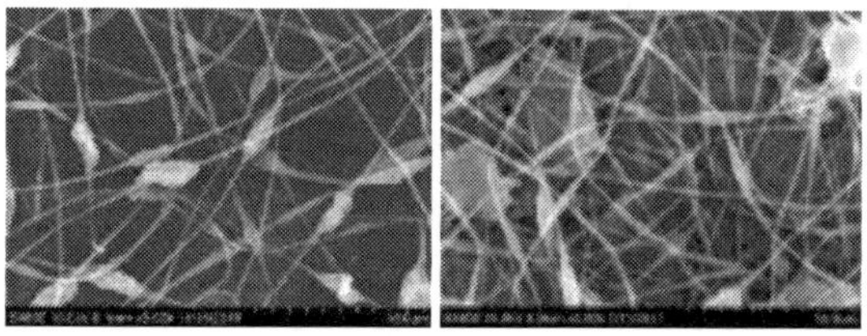

Figure 2. SEM images of PMMA fibers from PMMA solutions: (left) M_w~120,000, 8 wt%; (right) M_w~400,000, 1 wt%.

Figure 3. SEM image of PMMA fibers from the 1.5 wt% PMMA (M_w~400,000) solution.

Because electrospinning is operated by counteracting of electrostatic repulsion to surface tension, conductivity of the solvent is also a significant factor. Thus the effect of water quality on the fiber diameter was studied. Conductivities of distilled water and deionized water were 1.2 and 1.92 μS/cm, respectively. When the flow rate and voltage were kept at 0.05 mL/min and 7.1 kV, the average diameters of the formed fibers were 770 ± 45 nm (distilled water) and 650 ± 46 nm (deionized water). A higher conductivity of a solvent equals to more charge density. As the charges are introduced by applying high voltage, the repulsion of charges will stretch the jet, which causes narrower fibers. The effect of voltage on the fiber diameter was also studied. The average diameters were 837 ± 139 and 770 ± 45 nm for 5.5 and 7.1 kV, respectively. A high voltage gave a small average diameter and also a narrow diameter distribution. The result showed that the required diameters for different applications can be controlled by changing voltage.

PMMA is a hydrophobic polymer. Although it was dissolved in rubbing alcohol and then spun, the characteristic of hydrophobicity was not changed. Through the contact angle test, the mats achieved water contact angles > 130 °. The water-resistant property was also tested. A drop of tap water was dropped on the mat surface, and kept it for 10 min in order to understand the water resistance of the form fibers. The morphologies of the fibers before and after wetting were observed by SEM (Figure 4). After wetting, the fibers were not dissolved in water. Besides, there was not remarkable damage or deformation of the fibers.

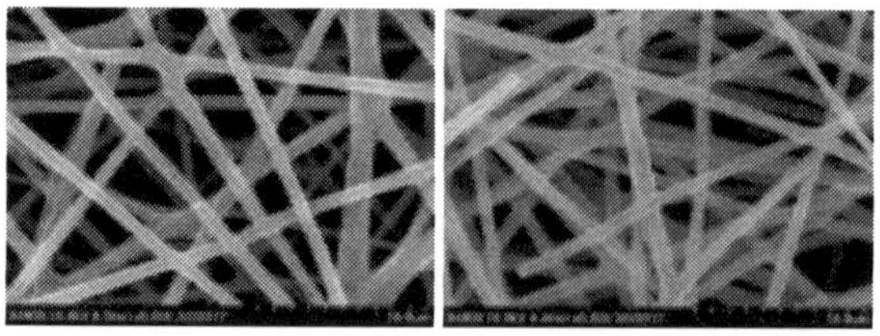

Figure 4. SEM images of PMMA fibers before (left) and after (right) wetting.

4. Conclusion

In this study, preparation of hydrophobic PMMA submicrofibers by electrospinning was demonstrated. An environmentally friendly solvent, i.e., rubbing alcohol (2-propanol/water = 7.8/2), was used to dissolve PMMA, and the solution was electrospun by controlling several parameters. Due to the high volatilization of the solution, PMMA with M_w~120,000 cannot be electrospun successfully. However, beads-free fibers can be formed when the concentration of PMMA with M_w~400,000 reached 1.5 wt%. Raising the voltage in electrospinning or conductivity of the solution will decrease the average diameter of the formed fibers. The PMMA mats were hydrophobic with contact angles > 130 ° and showed good water resistance.

5. References

1. N. Bhardwaj, S.C. Kundu, Biotechnol. Adv., 28, 325–347 (2010)
2. C.J. Luo, S.D. Stoyanov, E. Stride, E. Pelan, M. Edirisinghe, Chem. Soc. Rev., 41, 4708–4735 (2012)
3. S. Piperno, L. Lozzi, R. Rastelli, M. Passacantando, S. Santucci, Appl. Surf. Sci., 252, 5583–5586 (2006)
4. Y.F. Qian, Y. Su, X.Q. Li, H.S. Wang, C.L. He, Iran. Polym. J., 19, 123–129 (2010)
5. Y. Liu, Y. Ji, K. Ghosh, R.A.F. Clark, L. Huang, M.H. Rafailovichz, J. Biomed. Mater. Res., Part A, 90A, 1092–1106 (2009)
6. J.J. Miao, R.C. Pangule, E.E. Paskaleva, E.E. Hwang, R.S. Kane, R.J. Linhardt, J.S. Dordick, Biomaterials, 32, 9557–9567 (2011)
7. L.P. Cheng, H.Y. Shaw, J. Polym. Sci., Part B: Polym. Phys., 38, 747–754 (2000)
8. S.L. Shenoy, W.D. Bates, H.L. Frisch, G.E. Wnek, Polymer, 46, 3372–3384 (2005)
9. S.L. Shenoy, W.D. Bates, G. Wnek, Polymer, 46, 8990–9004 (2005)

Simplified shear relaxation modulus function of HDPE for FE simulation in extrusion blowing process

Nattarawee Siripath[1,a] and Satjarthip Thusneyapan[1]

[1]*Department of Mechanical Engineering, Faculty of Engineering, Kasetsart University, Bangkok, 10900, Thailand.*

Abstract. Attempts to apply finite element analysis for simulation of the final geometry from extrusion blowing process have been studied. One of the accuracies for the prediction is a suitable mathematical model of the thermoplastic during the process. Visco-elastic model is found to be a suitable model by many researchers. This paper simplified the shear relaxation modulus function (SRMF) which characterizes the high-density polyethylene in molten state. Instead of the series of five exponential functions previously used, our simplified-SRMF consists of one power function of temperature and time. The deviation of our model from Williams-Landel-Ferry model with KBKZ constants, between the temperature of 140°C to 200°C and time from 0 to 10 sec, is 9.3706%. When simulated on a bottle, our model has the final bottle thickness and time deviation from WLF-KBKZ model by 0.19622% and 3.07362% respectively.

1 Introduction

Research in employing finite element analysis (FEA) technology for simulation the extrusion blow molding process have been performed by many scholars [1-6]. The objective of the analysis is to find the wall thickness of products [2, 3, 5, 6]. The thickness depends on the process conditions – mainly the parison conditions. The parison is a long hollow thermoplastic molten tube. Air pressure blows inside the parison; thus, expands the tube against the mold cavity surface to form the shape of products. The high-density polyethylene (HDPE) uses in extrusion blow molding process; its hot and soft parison exhibits visco-elasic characteristic. Hence, the isotropic visco-elastic model is good for modelling the deformation of HDPE parison in molten state [5, 6]. The material property for the process simulation is shear relaxation modulus function or SRMF [4, 5, 6]. The time-temperature shift function introduced by Williams-Landel-Ferry (WLF) is applied in SRMF [7]. The procedure of applying WLF together with Kaye-Bernstein-Kearsley-Zapas (KBKZ) material constants introduces a series of five exponential functions [4, 5, 6]. This paper introduces a simplified SRMF as a function of temperature and time; it has only one power function.

2 Mathematical model for HDPE

Visco-elasticity is the property of HDPE, it represents both viscosity of fluid and elasticity of solid. In mechanical components, a viscous damper characterizes the viscosity, and an elastic spring is the elasticity. The mathematical model of the visco-elastic for finite element (FE) simulation considers shear stress and shear strain [8]. A linear elastic material with shear elastic modulus of G, according to Hooke's law, shear stress of the elastic element τ_e relates to its shear strain γ_e by

$$\tau_e = G\gamma_e. \tag{1}$$

The shear stress of the viscous material τ_v is characterized by the shear strain rate $d\gamma_v/dt$ as

$$\tau_v = \eta\frac{d\gamma_v}{dt}, \tag{2}$$

where η is the viscosity and γ_v is the shear strain in the damper (or viscous element).

Two basic visco-elastic models are Kelvin-Voigt and Maxwell. Since the deformation of the parison has more of viscous characteristic than elastic, the Maxwell material model is suitable for the application. The Maxwell model configuration has a damper in series with a spring. Under this circumstance, the total shear strain is

$$\gamma = \gamma_e + \gamma_v. \tag{3}$$

For a constant area, shear stress in the material τ equals to the shear stress of both elastic τ_e and viscous τ_v or

$$\tau = \tau_e = \tau_v. \tag{4}$$

The shear strain rate of the elastic element is found by differentiating Eq. (1), that is

$$\frac{d\tau_e}{dt} = G\frac{d\gamma_e}{dt}.$$

Since $\tau = \tau_e$ and from this equation, the shear strain rate of the elastic element is

$$\frac{d\gamma_e}{dt} = \frac{1}{G}\cdot\frac{d\tau}{dt}. \tag{5}$$

[a] Corresponding author: s.natwee@gmail.com

As $\tau = \tau_v$, and solving for the shear strain rate of Eq. (2), we have

$$\frac{d\gamma_v}{dt} = \frac{\tau}{\eta}. \tag{6}$$

By taking the derivative of Eq. (3), we obtain

$$\frac{d\gamma}{dt} = \frac{d\gamma_e}{dt} + \frac{d\gamma_v}{dt}. \tag{7}$$

Substituting Eqs. (5) and (6) into Eq. (7), the total shear strain rate becomes

$$\frac{d\gamma}{dt} = \frac{1}{G}\cdot\frac{d\tau}{dt} + \frac{\tau}{\eta},$$

which is a first-order differential equation of the form

$$\dot{\tau} + \frac{G}{\eta}\tau = G\dot{\gamma}. \tag{8}$$

In the stress relaxation test, the strain is constant; then, strain rate $\dot{\gamma}$ is zero. Equation (8) becomes

$$\dot{\tau} + \frac{G}{\eta}\tau = 0. \tag{9}$$

The solution of the homogeneous differential equation of Eq. (9) is

$$\tau(t) = \tau_o \exp\left(-\frac{t}{\Gamma}\right). \tag{10}$$

where τ_o is the initial shear stress at time $t = 0$, or $\tau_o = \tau(0)$, and

$$\Gamma = \frac{\eta}{G}. \tag{11}$$

The constant Γ is the shear stress relaxation time; it is the ratio of the viscosity η and the shear elastic modulus G which has the unit of time. The shear stress relaxation time indicates the time taken the shear stress to relax or reaches zero; it is a dynamics property of the material.

Dividing Eq. (10) through by γ, we obtain the shear modulus as a function of time as

$$G(t) = G_o \exp\left(-\frac{t}{\Gamma}\right). \tag{12}$$

For a material with several Maxwell models in parallel, each with viscosity η_i and the shear stress relaxation time Γ_i, the shear stress relaxation modulus function (SRMF) is a Prony series as

$$G(t) = \sum_{i=1}^{n} G_i \exp\left(-\frac{t}{\Gamma_i}\right). \tag{13}$$

For thermoplastic materials, the effect of temperature is described by time-temperature shift function $w(T)$ proposed by Williams-Landel-Ferry [4, 7]. The function $w(T)$ is known as WLF model, and is defined by

$$w(T) = \exp\left(\frac{-c_1(T - T_{ref})}{c_2 + (T - T_{ref})}\right). \tag{14}$$

The parameters of the function $w(T)$ are temperature coefficient c_1, temperature constant c_2 and the reference temperature T_{ref} in °C. These constants for HDPE for the WLF function are in Table 1 [4]. Using the data in Table 1, time-temperature shift function $w(T)$ for HDPE is

Table 1. The HDPE constants of WLF function.

Parameter	Value
c_1	6.928
c_2	350°C
T_{ref}	150°C

$$w(T) = \exp\left(\frac{-6.928\,(T - 150)}{350 + (T - 150)}\right). \tag{15}$$

Notice that the high temperature T reduces the value of the function $w(T)$. During the blowing process, the molten HDPE temperature is vary, and causes the change in the SRMF. This temperature affects the SRMF by introducing $w(T)$ as a multiplier of Γ_i as shown in Eq. (16), and it becomes a function of two variables: temperature T and time t. This time t is referred to relaxation time.

$$G(T, t) = \sum_{i=1}^{n} G_i \exp\left(-\frac{t}{\Gamma_i w(T)}\right). \tag{16}$$

The value of Γ_i and G_i of HDPE at 150°C according to the KBKZ material constants are shown in Table 2 [4].

Table 2. The KBKZ constants Γ_i and G_i of HDPE at 150°C.

i	Γ_i (sec)	G_i (MPa)
1	0.03906	0.2475
2	0.3125	0.07243
3	2.5	0.03046
4	20	0.007814
5	160	0.003188

The SRMF or $G(T, t)$ of HDPE of Eq. (16) by using data of Γ_i and G_i in Table 2 is

$$G(T,t) = 0.2457\exp\left(-\frac{t}{0.03906\,w(T)}\right) + 0.07243\exp\left(-\frac{t}{0.3125\,w(T)}\right) + 0.03046\exp\left(-\frac{t}{2.5\,w(T)}\right) + 0.007814\exp\left(-\frac{t}{20\,w(T)}\right) + 0.003188\exp\left(-\frac{t}{160\,w(T)}\right). \tag{17}$$

For the temperatures from 140°C to 200°C and relaxation times t from 0.01 to 100 seconds, the values of the function $G(T, t)$ is plotted in Figure 1; some selected values are tabulated in Table 3.

3 Mathematical model of the simplified-SRMF

The SRMF of Eq. (17) is a series of exponential functions and it is closed to be linear on log-log scale as shown in Figure 2a. With this characteristic, the SRMF can be expressed by a simple power function as

$$G^*(T, t) = at^b, \tag{18}$$

where a is the coefficient of t, it is also a function of the temperature, or $a = a(T)$, and b is the power of t, and is a function of the temperature, or $b = b(T)$.

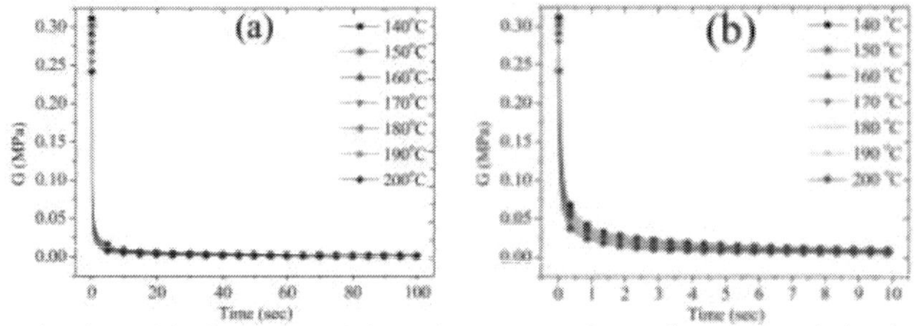

Figure 1. The shear relaxation modulus function $G(T, t)$ plots at the temperature from 140°C to 200°C and relaxation time from: (a) 0 to 100 seconds, and (b) a close-up at 0 to 10 seconds.

Table 3. Tabulation of SRMF of Eq. (17), at five selected times t from 0.01 to 100 sec, and the temperature from 140°C to 200°C with an increment of 10°C.

T (°C)	$G(T, t)$, MPa				
	$i = 1$ $t = 0.01$ sec	$i = 2$ $t = 0.1$ sec	$i = 3$ $t = 1$ sec	$i = 4$ $t = 10$ sec	$i = 5$ $t = 100$ sec
140°C	0.311318928	0.12668616	0.037979565	0.009392852	0.002047137
150°C	0.301688795	0.11181232	0.033971399	0.008292174	0.001759064
160°C	0.291129554	0.100143051	0.030771846	0.007456435	0.001512679
170°C	0.279737765	0.091096553	0.028140791	0.006778176	0.001290089
180°C	0.267641012	0.083967795	0.025868246	0.006185415	0.001084033
190°C	0.254993521	0.078103762	0.023808775	0.005640652	0.000893834
200°C	0.241969952	0.073007628	0.021884103	0.005130375	0.00072152

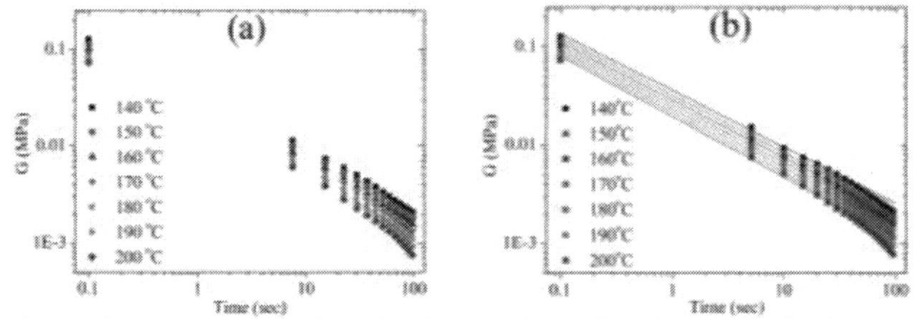

Figure 2. (a) Plot of $G(T, t)$ from Figure 1 on a log-log scale. (b) Comparing the linear least squares functions (21) to (27) with $G(T, t)$ at the 7 temperatures and the time to 100 sec.

Taking the natural logarithm on both sides of the equation, we have

$$\ln [G^*(T, t)] = \ln(at^b),$$

or

$$\ln [G^*(T, t)] = \ln(a) + b\ln(t). \tag{19}$$

Eq. (19) is a linear function when we write it as

$$G^*_{\ln} = A + bt_{\ln}, \tag{20}$$

where $G^*_{\ln} = \ln [G^*(T, t)]$, $A = \ln(a)$ and $t_{\ln} = \ln(t)$.

By using the linear least squares fitting, we obtain the SRMF at each temperature from 140°C to 200°C as follows.

$$G^*(140°C, t) = 0.03619t^{-0.57512}, \tag{21}$$
$$G^*(150°C, t) = 0.03223t^{-0.58048}, \tag{22}$$
$$G^*(160°C, t) = 0.02890t^{-0.58644}, \tag{23}$$
$$G^*(170°C, t) = 0.02607t^{-0.59296}, \tag{24}$$
$$G^*(180°C, t) = 0.02363t^{-0.59990}, \tag{25}$$
$$G^*(190°C, t) = 0.02151t^{-0.60713}, \tag{26}$$
$$G^*(200°C, t) = 0.01963t^{-0.61453}. \tag{27}$$

The graph of Eqs. (21) to (27) from time of 0 to 100 seconds with the superimposed data points from Figure 2a are shown in Figure 2b.

The coefficient a of Eqs. (21) to (27) is plotted against the temperature as shown in Figure 3a; between 140°C to 200°C. The coefficient a is shown to be linear with the temperature T. The variation of a is approximated by a linear function

$$a(T) = A_a + B_a T. \tag{28}$$

By using the linear least squares method, it yields.

$$a(T) = 0.07858 - 0.000297214\,T, \tag{29}$$

For positive value of $G^*(T, t)$, the coefficient $a(T)$ must be positive; hence, the condition for the temperature is $T < 246.34$°C. The goodness of fitting, of Eq. (29) between 140°C and 200°C, consists of the coefficient of multiple determination R^2 and the standard error of the estimate σ_e; they are $R^2 = 0.98181$ and $\sigma_e = 0.005252$. The R^2 value is almost one, and σ_e is near zero, they

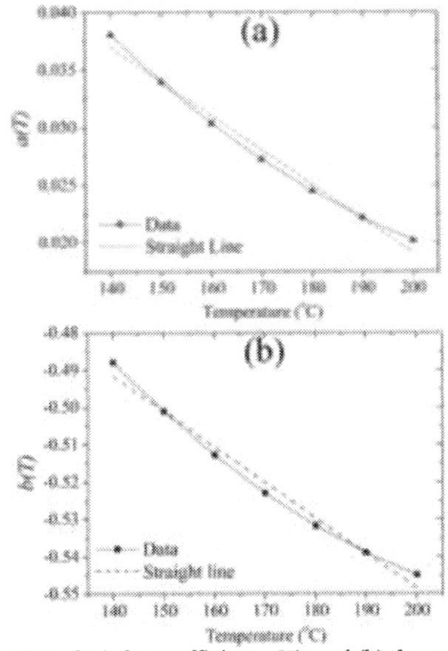

Figure 3. Plot of (a) the coefficient $a(T)$, and (b) the power $b(T)$. The straight lines are their linear approximation of functions (29) and (30) respectively.

indicate the good curve fitting of the linear function to the data.

Similarly, the power b of Eqs. (21) to (27) is observed to be linear between 140°C and 200°C, Figure 3b. The linear characteristic of b is a function of the temperature T; by using the linear least squares calculation gives

$$b(T) = -(0.35868 + 0.0009493574T). \quad (30)$$

This function has $R^2 = 0.97946$ and $\sigma_e = 0.016777$. These two values indicate good fitting results.

Finally, the simplified-SRMF under the reference temperature at 150°C is.

$$G^*(T, t) = (0.07858 - 0.00030T)\, t^{-(0.35868 + 0.00095T)}. \quad (31)$$

Referring to Table 3, this function $G^*(T, t)$ is recommended for 140°C $\le T \le$ 200°C and $0 \le t \le 100$ sec.

The function $G^*(T, t)$ in Eq. (31) is our simplified-SRMF (or S-SRMF) for the HDPE. The deviations of $G^*(T, t)$ from $G(T, t)$ vs. time of the 7 temperatures are shown in Figure 4a. The average of the 7 temperatures of the absolute deviation $|\Delta_T\%|$ at each time is shown in Figure 4b. The average deviations $|\Delta_T\%|$ against the temperature at 140°C to 200°C for the time from 0.01 to 10 sec and 0.01 to 100 sec are shown in Figure 4c; the overall averages are 9.3706% and 27.3335%, respectively. Even though the deviation at 100 sec is large, it has no great effect to the deviation; since most of the blow molding time is completed in less than 2 seconds.

4 The use of S-SRMF for FEA thickness analysis

In order to examine the accuracy of the S-SRMF in the FE simulation of the blowing process, a square bottle with the width, length and height of 76, 76 and 170 mm

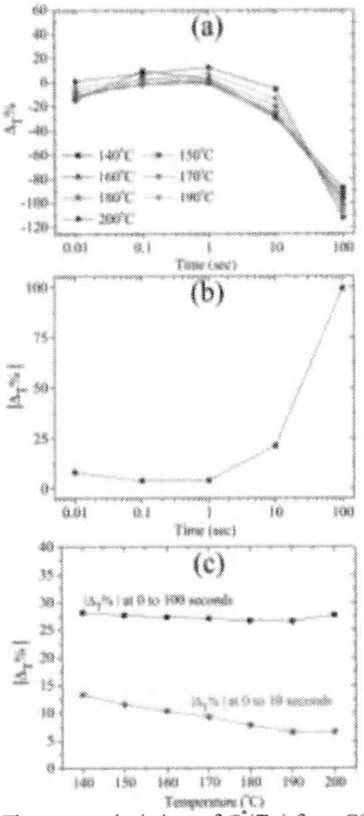

Figure 4. The percent deviations of $G^*(T, t)$ from $G(T, t)$. (a) $\Delta_T\%$ vs. T at t = 0.01 to 100 sec. (b) Average of the absolute deviation $|\Delta_T\%|$ vs. t of 7 temperatures. (c) Average of $|\Delta_T\%|$ vs. T for time from 0.01 to 10 sec and 0.01 to 100 sec.

was selected (Figure 5a). The FE model (FEM) for the parison, with the diameter of 38.5 mm and length of 185 mm, has 3504 nodes and 3456 elements (Figure 5b). Its quadrilateral element size is 2.5×2.5 mm and 2.2 mm thick.

The isotropic visco-elastic model using WLF-KBKZ model ($G(T, t)$ of Eq. (17)) and our S-SRMF, $G^*(T, t)$, of Eq. (31) were assigned as the material of the parison elements. The processing condition was set according to Rugsaj [6]. The FEA program for the simulation was MSC. MARC 2005. Figure 5c is an illustration of the deforming parison at time 0.5, 1.0, 1.1, 1.2 and 1.5 seconds when using $G^*(T, t)$; note that the deform for $G(T, t)$ has similar shape.

5 Thickness deviation

The comparison of the thickness was done at 0°, 45° and 90° from the parting line (PL) and at 20% to 70% of the bottle height, as shown in Figure 6a. The graph in Figure 6b compares the thickness, at the mentioned points, of $G^*(T, t)$ and $G(T, t)$. The thickness deviation between the two SRMFs in Figure 6b has the maximum deviation of 0.85742%. The average deviation along the 20% to 70% of the height at 0°, 45° and 90° from PL are 0.05258%, 0.38358% and 0.05005% respectively.

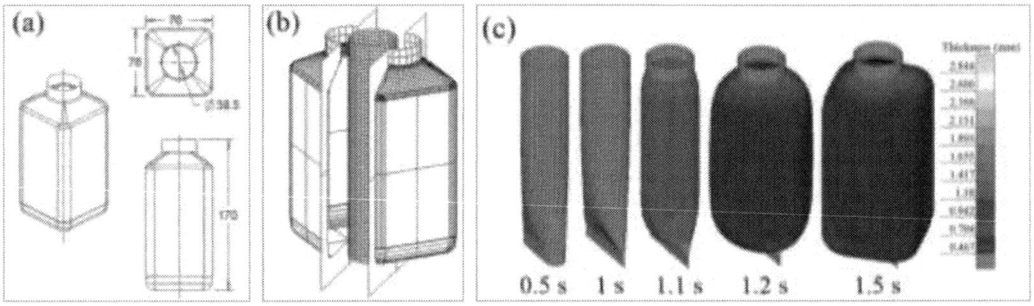

Figure 5. (a) Bottle dimension. (b) FEM of process. (c) Deforming parison at 0.5 to 1.5 sec.

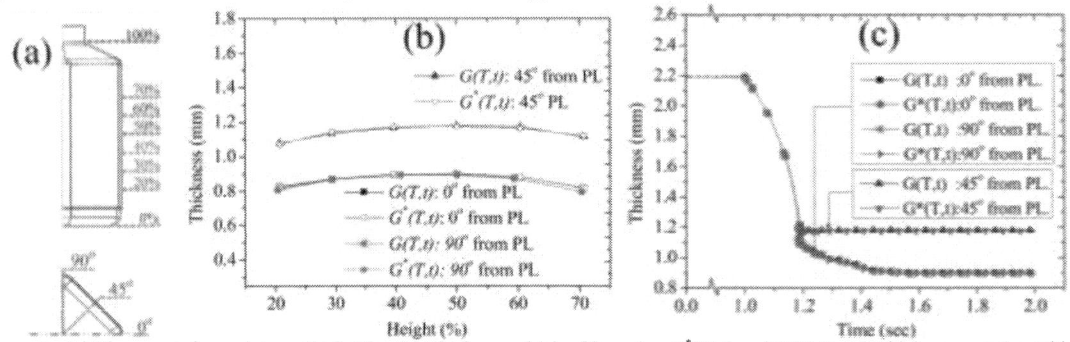

Figure 6. (a) The measuring points on the bottle. (b) Thickness obtained by using $G^*(T, t)$ and $G(T, t)$. (c) Thickness varying with time at 0°, 45°, 90° and 50% height of $G^*(T, t)$ and $G^*(T, t)$.

The thickness varying with the blowing time at 0°, 45° and 90° from PL and 50% height is shown in Figure 6c. The thickness responses of the model when using $G^*(T, t)$ and $G(T, t)$ look identical. The time deviations when the thicknesses become 1.2 mm, at 0°, 45° and 90° from PL are 0.03653, 0.03653 and 0.03653 msec, respectively; the average of this time difference is 0.03653 msec or 3.073622%.

The final thickness at 2 sec of Figure 6c of the three points has the average deviation of 0.79686%. The thicknesses are observed to begin to have constant thickness at time 1.2 and 1.5 sec. The results in these graphs indicated the compatibility of the S-SRMF with the model using WLF-KBKZ data.

6 Conclusion

The popularly used visco-elastic material model for extrusion blow molding for finite element simulation is based on WLF model and KBKZ data. Their shear relaxation modulus function or SRMF for HDPE contain a sum of five exponential functions of temperature and time. This paper presents a simplified-SRMF (S-SRMF) which has only one power function. The S-SRMF or $G^*(T, t)$ is numerically obtained from the data of WLF-KBKZ from the temperature at 140°C to 200°C, and time from 0 to 100 sec. Our S-SRMF is found easier for calculation and suitable for FE analysis. It has the average deviation from WLF-KBKZ data of 9.3706% for time from 0 to 10 sec. The FE simulation of a square bottle, our S-SRMF when compared with WLF-KBKZ model, has the final bottle thickness deviation of 0.19622% and the average time deviation when the thickness become 1.2 mm of 3.07362%; therefore, the S-SRMF gives a comparable result with the conventionally used SRMF for both thickness and time response.

References

1. S.I. Tanifuji, *Process Poly. Eng. Sci.*, **40**, 16 (2008)
2. K. Peplinski, *Journal of Polish CIMAC*, **5(3)**, 6 (2010)
3. K. Peplinski, *Journal of Polish CIMAC*, **7(3)**, 6, (2012)
4. D. Laroche, K.K. Kabanemi, L. Pecora, R.W. Diraddo, *Poly. Eng. Sci.* **39(7)**, 11 (1999)
5. S. Thusneyapan, R. Rugsaj, *Journal of Research App. Mech. Eng.*, **4(1)**, 12-24 (2016)
6. R. Rugsaj, ME Thesis (Bangkok, Kasetsart University, 2014)
7. M.L. Williams, R.F. Landel, J.D. Ferry, *J. Amer. Chem. Soc.* **77**, 7 (1955)
8. D.V. Rosato, A.V. Rosato, D.P. Dimattia, *Blow Molding Handbook (2nd revised edition)*, Chapter 7 (2003)

The Synthesis and Characterization of a Novel Cationic Asphalt Emulsifier

Xue-Di Gong, Lai-Shun Shi[a], Tian-Yao Wang, Xiao-Meng Yu

School of Chemistry and Chemical Engineering, Shandong University, Jinan 250061, P. R. China

Abstract. A new cationic asphalt emulsifier of N,N-dimethyl-N-(3-(N',N'-dimethyl amido)-2-hydroxy propyl)-coconut oil amide propyl-1-ammonium chloride was synthesized by two steps reaction of coconut oil acyl propyl dimethyl tertiary amine (PKO), epoxy chloropropane and dimethylamine. The chemical structure of the key intermediate of N,N-dimethyl-N-(ethylene oxide-2-methylene)-coconut oil amide propyl-1-ammonium chloride was confirmed by FTIR, ^{1}H NMR and elemental analysis. The optimum reaction condition of first step was obtained by single factor analysis: reaction time 5 h, reaction temperature 50 ℃, feedstock mole ratio of epoxy chloropropane to PKO 1.05. The reaction yield is 82.15% and the epoxy value is 40.39% at the optimum conditions. The critical micelle concentration (CMC) of the asphalt emulsifier is 7.80×10^{-2} mol/L. The surface tension at CMC is 24.57 mN/m. The emulsifier showed excellent emulsification effect for the asphalt. The prepared bituminous emulsion had higher storage stability. The emulsifier belongs to medium-set asphalt emulsifier.

1 Introduction

Emulsified asphalt is composed of asphalt, emulsifier, water and other additives. Compared with the hot asphalt, emulsified asphalt can be applied at room temperature, has the advantages of better adhesive ability with stone, low energy consumption and less environmental pollution. It is widely used in highway and waterproof materials [1,2].

Emulsified asphalt is an oil-in-water type emulsion. Asphalt emulsifier can be classified into cationic, anionic, amphoteric and non-ionic [3,4]. At the beginning, the mainly used asphalt emulsifier was anionic, such as long chain sulfonate, the alkali metal salt of lignin, and carboxylic acid salt, etc [5]. Cationic emulsifier has developed rapidly in recent years [6,7]. Cationic asphalt emulsifier has good emulsification performance and strong adhesion force with stone. Most of the stone has a negative charge. Cationic emulsified asphalt can rapid adsorb and bind on the surface of stone. The mainly cationic emulsifiers are nitrogen containing organic amine derivatives and quaternary ammonium salt. Li *et al* [8] synthesized an asymmetric Gemini cationic surfactants, characterized its structure and discussed the affect factors of stability. Banno *et al* [9] synthesized novel cationic surfactants which have biodegradability and chemical recyclability green sustainable.

In previous papers, different kinds of asphalt emulsifiers were synthesized by our research group [7,10-14]. A new cationic asphalt emulsifier of N,N-dimethyl-N-(3-(N',N'-dimethyl amido)-2-hydroxy propyl)-coconut oil amide propyl-1-ammonium chloride was synthesized by two steps reaction of coconut oil acyl propyl dimethyl tertiary amine (PKO), epoxy chloropropane and dimethylamine in this paper. The structure of the emulsifier was confirmed by FTIR, ^{1}H NMR and elemental analysis. The emulsifier belongs to medium-set asphalt emulsifier.

2 Experimental

2.1 Materials and characterization

Coconut oil acyl propyl dimethyl tertiary amine (PKO) was technical grade and obtained from Lianyungang Yihai Chemical Co., Ltd, Lianyungang, China. The other chemical reagents were analytical reagent grade. The asphalt used was AH-90, which was provided by Dongming Petrochemical Company, Dongming, Shandong, China. The mineral aggregate was the mixture of different size particles of marble stone. The FTIR spectrum was measured at 400 to 4000 cm^{-1} wavenumber range with averaging 16 scans at a resolution of 4 cm^{-1} on a Tensor-27 FTIR spectrophotometer (Bruker, Germany). The ^{1}H NMR spectrum was measured on a Bruker Avance 300M NMR spectrometer (Bruker, Germany), using $CDCl_3$ as solvent and TMS as internal reference. Chemical shifts (δ) were given in ppm. Surface tension of asphalt emulsifier solutions was measured with Krüss Processor Tensiometer K12 (Krüss GmbH, Hamburg, Germany). Elemental analysis was measured by the Elementar vario EL type III element analyzer (Elementar, German).

[a] Corresponding author: LSHUNSH@sdu.edu.cn

2.2 Synthesis of asphalt emulsifier

Scheme 1 shows the synthesis step of the cationic asphalt emulsifier.

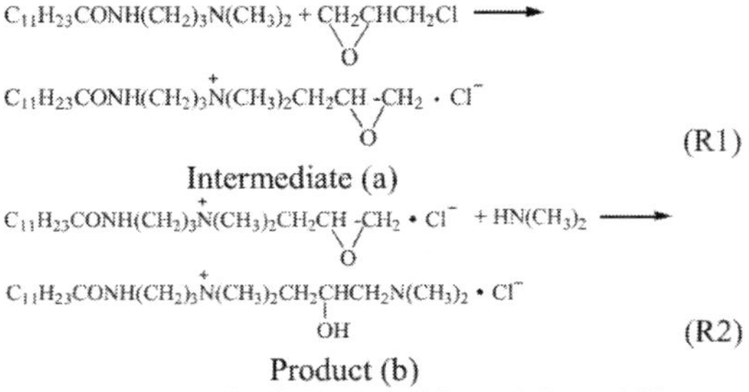

Scheme 1. Synthetic step of the asphalt emulsifier.

N,N-dimethyl-N-(ethylene oxide-2-methylene)-coconut oil amide propyl-1-ammonium chloride (a): Coconut oil acyl propyl dimethyl tertiary amine (14.225 g, 0.05 mol) was dissolved in 16 mL anhydrous ethanol. Then, epoxy chloropropane (4.856 g, 0.0525 mol) was added in it. The mixture was stirred for 5 h at 50℃ in a water bath. The solvent in the mixed solution was removed by vacuum distillation. The intermediate (a) was recrystallized by anhydrous diethyl ether for three times, and dried in vacuum for 12 hours.

N,N-dimethyl-N-(3-(N',N'-dimethyl amido)-2-hydroxy propyl)-coconut oil amide propyl-1-ammonium chloride (b): 33% dimethylamine aqueous solution (17.1 g, 0.125 mol) was added in intermediate (a) synthesized above. The mixture was stirred for 3 h at 65 ℃ in a water bath. The solvent in the mixed solution was removed by vacuum distillation. The product (b) was recrystallized by acetone for three times, and dried in vacuum for 12 hours.

2.3 The yield of intermediate (a)

The yield (Y) of intermediate (a) was obtained by Mohr analysis method. For reaction (R1), chloride ion was generated. 1.5000 g sample was weighed into a conical flask, then 50 mL distilled water and 0.1 mL of 5% potassium chromate indicator was added. The solution was titrated with 0.1 mol/L $AgNO_3$ standard solution. The titration endpoint is the appearance of red-brown silver chromate. A blank titration was carried out at the same time.

$$Y = \frac{N_1 \times (V_1 - V_{01}) \times m}{1000 \times m_1 \times n_1} \times 100\% \qquad (1)$$

where, V_1 means the consumed volume of silver nitrate standard solution, mL. V_{01} means the consumed volume of silver nitrate standard solution in blank titration, mL. m means the total mass of sample, gram. m_1 means the mass of sample to be titrated, gram. N_1 means the concentration of silver nitrate standard solution, mol/L. n_1 means the mole of epoxy chloropropane, mol.

2.4 Epoxy group value

The epoxy group value (E) of intermediate (a) in reaction (R1) was measured according to the following method. 1.5000 g sample was weighed into a conical flask, then 10 mL concentrated hydrochloric acid (HCl)-acetone solution (volume ratio 1:40) and 10 mL ethanol were added. The solution was reflux at 70 ℃ for 1 h, and then was cooled to room temperature. 0.1 mL of phenolphthalein indicator was added. The solution was titrated with standard sodium hydroxide solution (0.1 mol/L). The titration endpoint was the appearance of pink color. A blank determination was carried out at the same time.

$$E = \frac{N_2 \times (V_2 - V_{02}) \times m}{1000 \times m_2 \times n_2} \times 100\% \qquad (2)$$

Where, V_2 means the consumed volume of sodium hydroxide standard solution, mL. V_{02} means the consumed volume of sodium hydroxide standard solution in blank determination, mL. m means the total mass of sample, gram. m_2 means the mass of the sample to be titrated, gram. N_2 means the concentration of sodium hydroxide standard solution, mol/L. n_2 means the mole of PKO, mol.

2.5 Surface tension and CMC

Surface tension of the asphalt emulsifier solutions decreases sharply with the increase of concentration, and nearly does not change when the concentration gets CMC. The critical micelle concentration and the surface tension at CMC were at the turning point of the curve of surface tension to logarithm of concentration.

2.6 Performance test of bituminous emulsion

2.6.1 Preparation of bituminous emulsion

15.0 g asphalt emulsifier was dissolved in 200 g water, then the solution was heated to 65 ℃. The pH value of the asphalt emulsifier aqueous solution was adjusted to 2.0 by adding dilute HCl (1:1). 300 g asphalt (AH-90) was heated to 125 ℃. The bituminous emulsion was obtained by mixing the asphalt emulsifier aqueous solution and the asphalt at a colloid mill for 1 min.

2.6.2 Mixing experiment

100 g Mineral aggregate, 1 g cement and 8 g water were mixed in a bowl. Then, 11 g bituminous emulsion was added, and the mixture was stirred at a speed of 60 revolutions per minute. The mixing time was measured by observation of the mixing appearance. The room temperature was 23.0 ℃.

3 Results and discussion

3.1 Synthesis process optimization

3.1.1 Feedstock mole ratio

The reaction conditions were fixed at: reaction temperature 50 ℃, reaction time 5 h. Figure 1 gives the yield (Y) of intermediate (a) versus feedstock mole ratio of epichlorohydrin to coconut oil acyl propyl dimethyl tertiary amine (m). As shown in Fig. 1, the yield increases with the feedstock mole ratio (m). Then, the yield decreases as the feedstock mole ratio greater than 1.05. Figure 2 represents the epoxy group value (E) of intermediate (a) versus feedstock mole ratio of epichlorohydrin to coconut oil acyl propyl dimethyl tertiary amine (m). As shown in Fig. 2, the epoxy group value increases with the feedstock mole ratio (m). So, the optimum feedstock mole ratio (m) is selected as 1.05.

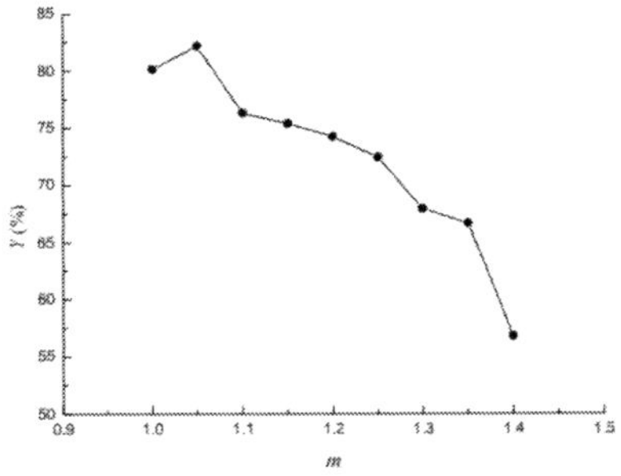

Figure 1. Yield vs. feedstock mole ratio.

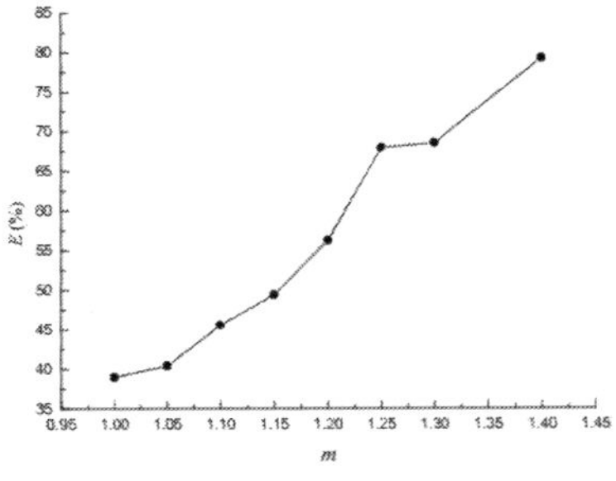

Figure 2. Epoxy group value vs. feedstock mole ratio.

3.1.2 Reaction time

The reaction conditions were fixed at: reaction temperature 50 ℃ , feedstock mole ratio of epichlorohydrin to coconut oil acyl propyl dimethyl tertiary amine 1.05. Figure 3 gives the yield (Y) versus reaction time (t). As shown in Fig. 3, the yield increases with the reaction time and gets maximum value at 6 h, then decreases with the reaction time. Figure 4 represents the epoxy group value (E) of intermediate (a) versus reaction time. As shown in Fig. 4, the epoxy group value increases with reaction time in the first stage, then decreases as the reaction time greater than 4 h. The decreasing of epoxy group value is of the reason of the hydrolysis and the ring opening of the epoxy group. So, the optimum reaction time is selected as 5 h.

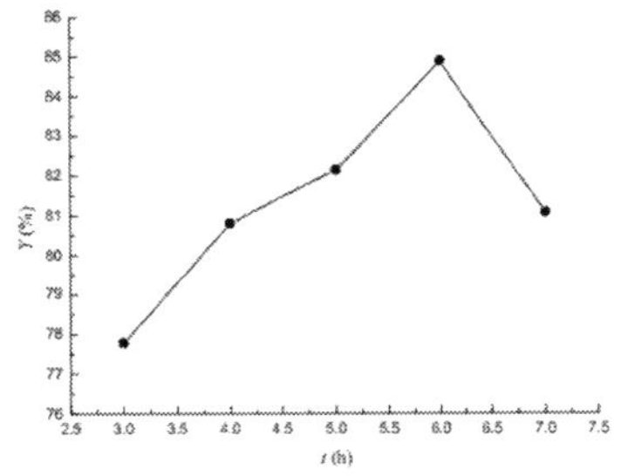

Figure 3. Yield vs. reaction time.

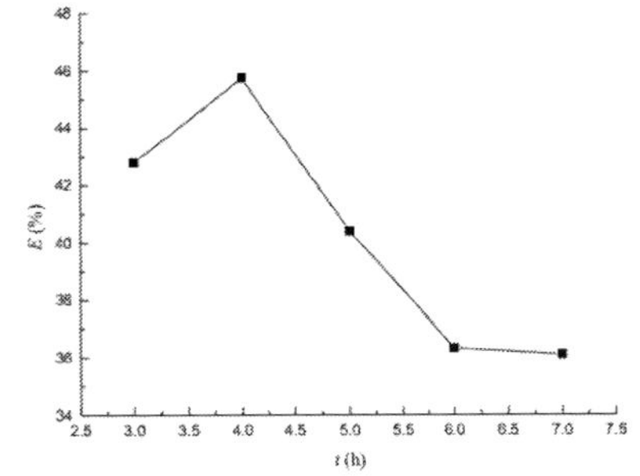

Figure 4. Epoxy group value vs. reaction time.

3.1.3 Reaction temperature

The reaction conditions were fixed at: reaction time 5 h, feedstock mole ratio of epichlorohydrin to coconut oil acyl propyl dimethyl tertiary amine 1.05. Figure 5 gives the yield (Y) versus reaction temperature (T). As shown in Fig. 5, the yield increases with the reaction temperature and gets maximum value at 60 ℃, then decreases with the reaction temperature. Figure 6 represents the epoxy group value (E) of intermediate (a) versus reaction temperature. As shown in Fig. 6, the epoxy group value decreases with the reaction temperature. The decreasing of epoxy group value is of the reason of the hydrolysis and the ring opening of the epoxy group at higher reaction temperature. So, the optimum reaction temperature is selected as 50 ℃.

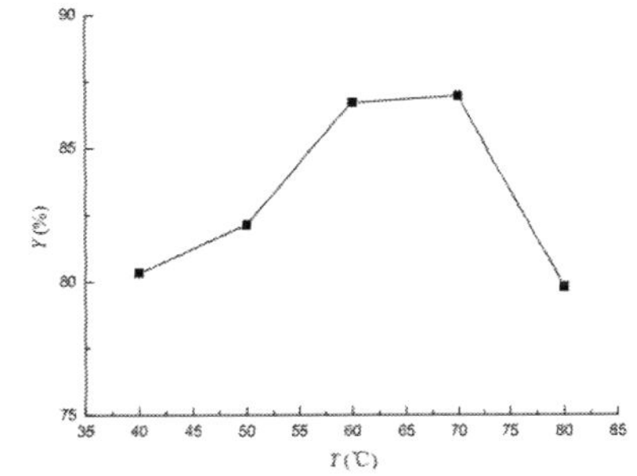

Figure 5. Yield vs. reaction temperature.

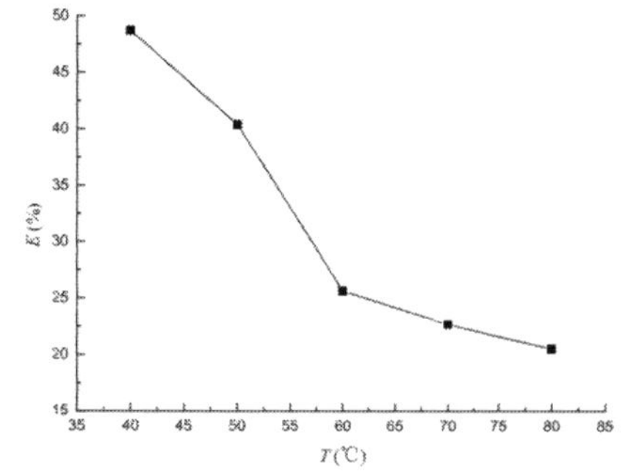

Figure 6. Epoxy group value vs. reaction temperature.

Therefore, the optimum reaction conditions can be got: reaction temperature 50 ℃, feedstock mole ratio of epichlorohydrin to coconut oil acyl propyl dimethyl tertiary amine 1.05, reaction time 5 h. The yield of intermediate (a) is 82.15%, the epoxy group value is 40.39% at the optimum reaction conditions.

3.2 FTIR characterization

Figure 7 is the FTIR spectrum of intermediate (a). The absorption at 3292 cm^{-1} (peak 1) originates from the N-H stretching in amide. The absorption at 3015 cm^{-1} (peak 2) originates from the C-H stretching in epoxy group. The absorption at 2800-3100 cm^{-1} originates from the C-H stretching. The absorption at 2920 cm^{-1} (peak 3) originates from the methyl asymmetrical stretching. The absorption at 2853 cm^{-1} (peak 4) originates from the methyl symmetrical stretching. The absorption at 1645 cm^{-1} (peak 5) originates from the C=O stretching in amide. The absorption at 1549 cm^{-1} (peak 6) originates from the N-H bending in amide. The absorption at 1466 cm^{-1} (peak 7) originates from the C-N stretching in amide. The absorption at 1260 cm^{-1} (peak 8) originates from the C-O stretching in epoxy group. The absorption at 964 cm^{-1} (peak 9) originates from the asymmetrical stretching of epoxy group. The absorption at 720 cm^{-1} (peak 10) originates from the CH_2 in-plane swing vibration. The absorption at 600 cm^{-1} (peak 11) originates from the N-H out-of-plane bending.

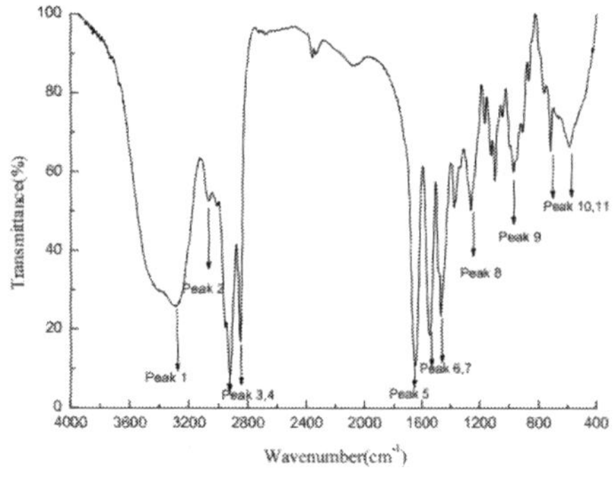

Figure 7. The FTIR spectrum of intermediate (a).

Figure 8 is the FTIR spectrum of asphalt emulsifier (b). The absorption at 3600-3216 cm^{-1} (peak 1) originates from the O-H stretching and N-H asymmetrical stretching. The absorption at 3078 cm^{-1} (peak 2) originates from the N-H stretching in amide. The absorption at 2928 cm^{-1} (peak 3) originates from the methyl asymmetrical stretching. The absorption at 2852 cm^{-1} (peak 4) originates from the methyl symmetrical stretching. The absorption at 1640 cm^{-1} (peak 5) originates from the C=O stretching in amide. The absorption at 1546 cm^{-1} (peak 6) originates from the N-H bending in amide. The absorption at 1460 cm^{-1} (peak 7) originates from the C-N stretching in amide. The absorption at 1362 cm^{-1} (peak 8) originates from the methylene symmetrical bending. The absorption at 1250 cm^{-1} (peak 9) originates from the N-H bending. The absorption at 1098 cm^{-1} (peak 10) originates from the C-O stretching. The absorption at 720 cm^{-1} (peak 11) originates from the CH_2 in-plane swing vibration. The absorption at 600 cm^{-1} (peak 12) originates from the N-H out-of-plane bending.

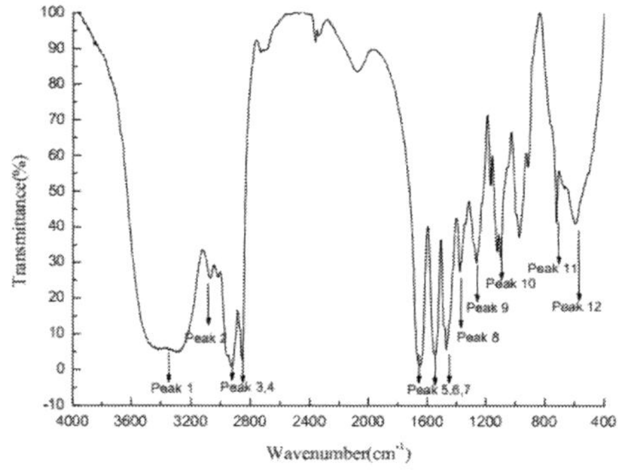

Figure 8. The FTIR spectrum of asphalt emulsifier (b).

3.3 ^{1}H NMR analysis

^{1}H NMR($CDCl_3$, 300Hz)δ:0.91(-$C\underline{H}_3$, t,3H); 1.32(-$(C\underline{H}_2)_8$-, m,16H); 1.57(C_9H_{19}-$C\underline{H}_2$-,s,2H); 2.05-2.20(-CONH-CH_2-$C\underline{H}_2$-, d,2H); 2.31($C_{10}H_{21}$-$C\underline{H}_2$-, t,2H); 3.15-3.70(-$C\underline{H}_2$-CH_2-$C\underline{H}_2$-N^+-$C\underline{H}_2$-$C\underline{H}$-$C\underline{H}_2$-, m,9H); 3.71-3.92(-N^+-$(CH_3)_2$, m,6H); 7.90-8.09(-NH-, m,1H) ppm. Therefore, the results of ^{1}H NMR analysis conform to the chemical structure of the intermediate (a).

3.4 Elemental analysis

Anal.: Calcd. for intermediate (a): C, 63.75%; H, 10.89%; N, 7.44%. Found: C, 57.00 %; H, 11.63 %; N, 6.94 %.

3.5 CMC and surface tension

Figure 9 gives the surface tension (σ) versus the logarithm of the asphalt emulsifier concentration (*c*) at 298 K. The surface tension showed a shape decrease firstly, then kept at a constant level. The CMC of the asphalt emulsifier is 7.80×10^{-2} mol/L. The surface tension at CMC is 24.57 mN/m.

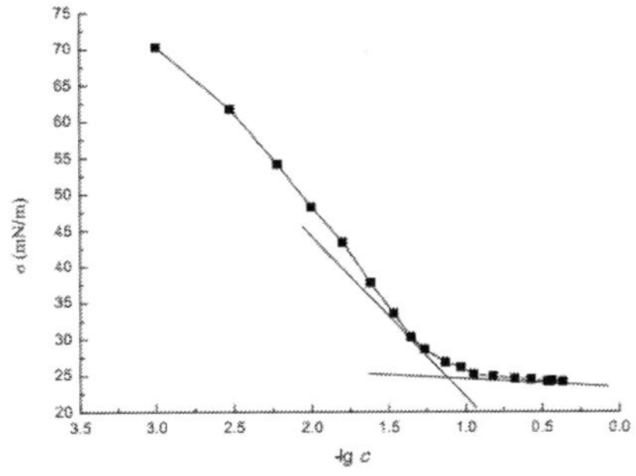

Figure 9. Surface tension vs. the logarithm of concentration.

3.6 Performance test of bituminous emulsion

The cationic asphalt emulsifier has good emulsifying ability for AH-90 asphalt. The bituminous emulsion to be prepared had higher storage stability. There was no separation of aqueous phase and asphalt phase appeared after 5 days. The mixing time was 60 s, which indicated that the emulsifier is a medium-set asphalt emulsifier. It can be applied to the application of tack coat, fog seal and chip seal in asphalt pavement construction.

4 Summary

A new cationic asphalt emulsifier was synthesized by coconut oil acyl propyl dimethyl tertiary amine (PKO), epoxy chloropropane and dimethylamine. The optimum reaction condition, the critical micelle concentration (CMC) and the surface tension at CMC of the asphalt emulsifier were obtained. The emulsifier belongs to a medium-set asphalt emulsifier. It can be applied to the application of tack coat, fog seal and chip seal in asphalt pavement construction.

Acknowledgement

The authors would like to thank the financial support from the Shandong Provincial Science and Technology Project (No. 2015GGX107005).

References

1. R.A. Mercado, J.L. Salager, V. Sadtler, et al. Breaking of a cationic amine oil-in-water emulsion by pH increasing: Rheological monitoring to modelize asphalt emulsion rupture. Colloids and Surfaces A: Physicochemical and Engineering Aspects, 458:63–68(2014)
2. F. Yang, G. Li, J. Qi, et al. Synthesis and surface activity properties of alkylphenol polyoxyethylene nonionic trimeric surfactants. Applied Surface Science, 257(1): 312-318(2010)
3. K. Isobe, R. Tamaki. Nonionic emulsifier for asphalt. U.S. Patent 6,114,418. (2000)
4. P.R.J. Blanpain, J.W. Creak, G.F. Mori, et al. Cationic asphalt emulsions. U.S. Patent 4,997,868. (1991)
5. W.A. McIntosh. Alkali lignin to stabilize slow-break asphalt emulsions. Industrial & Engineering Chemistry, 44(7): 1656-1659(1952)
6. N.A. Negm, M.A. El-Hashash, D.E. Mohamed, et al. Gemini cationic surfactants: synthesis and influence of chemical structure on the surface activity. Journal of Surfactants and Detergents, 16(5): 733-738(2013)
7. L. Shi, Y. Zhao. Synthesis and performance evaluation of a Gemini cationic asphalt emulsifier. Petroleum asphalt, 24(6): 66-71(2010)
8. H.P. Li, H. Zhao, K. Liao. The preparation of asphalt emulsions with dissymmetric Gemini quaternary ammonium salts cationic surfactants. Energy Sources, Part A: Recovery, Utilization, and Environmental Effects, 35(24): 2285-2293(2013)
9. T. Banno, K. Toshima, K. Kawada, et al. Synthesis and properties of gemini-type cationic surfactants containing carbonate linkages in the linker moiety directed toward green and sustainable chemistry. Journal of Surfactants and Detergents, 12(3): 249-259(2009)
10. L. Shi, M. Sun, N. Li, B. Zhang. A novel betaine type asphalt emulsifier synthesized and investigated by online FTIR spectrophotometric method. Chem. Ind. Chem. Eng. Q., 21, 113−121(2015)
11. X. Tao, L. Shi, M. Sun, N. Li. Synthesis of lignin amine asphalt emulsifier and its investigation by online FTIR spectrophotometry. Advanced Materials Research, 909, 72−76(2014)
12. N. Li, L. Shi, X. Gong, Q. Xu, X. Liu, X. Wang. Synthesis of a novel cationic asphalt emulsifier and its investigation by online FTIR spectrophotometry. Res. Chem. Intermed., 41(4), 1935−1950(2015)
13. C. Huai, L. Shi, N. Li. Synthesis of a novel betaine-type asphalt emulsifier and its investigation by online FTIR spectrophotometry. Research on Chemical Intermediates, 39, 597−614(2013)
14. Xiao-Meng Yu, Lai-Shun Shi, Cun-Fei Ma, Lu Hou, Jing-Wen Kong, Ya-Li Sun. The synthesis and characterization of novel bis-cationic asphalt emulsifier. MATEC Web of Conferences, 130: 04004, 80-84(2017)

AUTHOR INDEX